Mobility in the Face of

Extreme Hydrometeorological Events 2

Series Editor
Françoise Gaill

Mobility in the Face of Extreme Hydrometeorological Events 2

Analysis of Adaptation Rhythms

Edited by

Céline Lutoff
Séverine Durand

ISTE PRESS

ELSEVIER

First published 2020 in Great Britain and the United States by ISTE Press Ltd and Elsevier Ltd

ISTE Press Ltd
27-37 St George's Road
London SW19 4EU
UK

www.iste.co.uk

Elsevier Ltd
The Boulevard, Langford Lane
Kidlington, Oxford, OX5 1GB
UK

www.elsevier.com

British Library Cataloguing-in-Publication Data
A CIP record for this book is available from the British Library
Library of Congress Cataloging in Publication Data
A catalog record for this book is available from the Library of Congress
ISBN 978-1-78548-290-8

Contents

Preface

Living with and adapting to extreme events is becoming an imperative in the context of ongoing global warming. At the time of writing, we are experiencing a new heat wave in France with temperatures reaching 45°C in southeastern France. This region, as well as the entire Mediterranean, is also regularly affected by extreme hydrometeorological phenomena that are still too often the cause of many deaths and major damage. It is these phenomena of flash flooding that we propose to analyze here, by comparing hydrometeorological processes and social dynamics that contribute to increasing or reducing exposure at the time of the crisis or over a longer period of land use planning.

To question these interactions between extreme events and social response requires the collaboration of climate sciences and the human and social sciences. Several successive projects, all of which had southeast France or the Mediterranean region as their field of application, have contributed both to a better understanding of the processes involved and to the consolidation of our interdisciplinary approach.

The adventure began following the 2002 flood that affected the entire Gard department (southeast of France). Several research funding agencies, both public and private, have supported us in successive studies that have enabled us to take our approach further and to propose a summary here. The first support came from insurers and the MAIF Foundation, which financed and supported a first study to analyze deaths following flash floods [RUI 07a], [RUI 07b]. At the same time, our contribution to the *FloodSite*[1]

Preface written by Céline LUTOFF and Séverine DURAND.

1 http://www.floodsite.net.

project (European FP6 project) made it possible to consolidate certain results by comparing them with other work carried out elsewhere in Europe [CRE 09]; [RUI 08]. Subsequently, the *Agence Nationale de la Recherche*, ANR (French National Research Agency), supported several of the projects. The first, entitled *MedUp*[2], focuses on the place of uncertainties, their propagation in forecasting models and their consideration in decision-making processes in crisis situations [CRE 10], [CRE 13]. It was complemented by the Amphora project (*Interreg III B Medocc Project*), which focused on weather forecasts based on environmental risks and their use by decision-makers in a crisis context. The *AdaptFlood* project, which followed, targeted the processes of individual adaptation of daily mobility in hydrometeorological crisis situations. It contributed to the establishment of the *Water & Society*[3] summer school whose objective was to share our interdisciplinary experience with young international researchers on the theme of interactions between water and societies [RUI 12], [RUI 14]. The *PreDiFlood*[4] project focused on distributed flood forecasting for road management in the Cévennes-Vivarais region [AUB 14]. In particular, it has made it possible to develop a model for predicting road cuts in the event of flash floods in the Gard [NAU 12], a model that we are using to analyze the exposure of individuals on the road network. The *MobiClimEx*[5] project has broadened the issue of mobility in the face of flash floods, considering the impacts of floods at different time scales: with respect to daily mobility in short time frame and on residential mobility in the long term [DEB 16], [LUT 18], [SHA 16], [SHA 17]. Finally, since 2010, the MISTRALS program[6] has contributed to the financing of field missions and the valuation of work within the framework of the HyMEx[7] project, which aims to understand, quantify, and model the hydrological cycle in the Mediterranean basin and its evolution in a context of global warming, with particular attention to high impact events. The work is currently continuing through other ongoing projects (in particular *the ANR PICS*[8]). However, it seemed important to us at the end of the *MobiClimEx* project to take stock of the main progress and results achieved by these various projects.

---

2 https://www.umr-cnrm.fr/spip.php? Article167.

3 http://www.waterandsociety.net/.

4 http://heberge.lcpc.fr/prediflood/index.php.

5 https://anr.fr/Projet-ANR-12-SENV-0002.

6 http://www.mistrals-home.org/.

7 https://www.hymex.org/.

8 http://pics.ifsttar.fr/.

The first volume of this book, published in 2018 [LUT 18], proposed to define the reference scales to be used for the analysis of these specific interactions between flood dynamics and social responses. In the continuation of this first volume, we propose here to focus our attention on the rhythms of adaptation of individuals and collectives to extreme hydrometeorological phenomena. In the conclusion of the previous volume, we already mentioned the need to consider the question of adaptation rhythms. The challenge here is to understand what mobilizes the social response, on the one hand, and what drives it: how many floods does it take to see risk management measures implemented at the local level or more broadly? Should we wait until we have water up to our knees before agreeing to abandon our current activity and protect ourselves from floods?

As this question of the pace of adaptation can only be addressed by an interdisciplinary approach, we wanted to draw lessons from our experience and suggest that the reader enters this book through a journey in interdisciplinarity. We will then address the issue of adaptation rates by differentiating between the specific tempos of the crisis (short-term analysis) and those of the inter-crisis and prevention (long-term analysis).

Chapters 2 and 3 focus on this long-term analysis. Chapter 2 examines the rhythm of urbanization in flood-prone areas. Are they specific in relation to what can be observed elsewhere? Does the potential threat of flooding affect the collective choices and planning decisions of these particular areas? Chapter 3 focuses on individuals and questions the place of flooding in their residential choices. Through a qualitative approach conducted with residents and also with real estate and land professionals, it seeks to understand how extreme events affect an individual's decision to live or not live in areas subject to high water levels.

Chapters 4 and 5 focus on the rhythm of adaptation at the time of the event. They focus on the specific exposure of motorists in times of crisis. Based on an analysis of the daily mobility of the Gard populations and by crossing them with the knowledge acquired on road cuts, Chapter 4 proposes modeling of this exposure and tests it on the case of the Gard flood in 2002. Based on some of the conclusions of this chapter, the following section provides complementary elements to the analysis of daily mobility by using individuals' activity programs to refine the understanding of social daily rhythms and the place occupied by travel in them.

The last three chapters of the book examine the methodological aspects of these analyses of adaptation rates. Chapter 6 proposes to test the possible use of data from social networks as a tool for observing social reaction rates following an event. Chapter 7 proposes a reflection on the possible use of certain physical cursors to better understand the rhythms of adaptation of social mobility, both during the crisis and in the long term for prevention. Finally, Chapter 8 provides a summary of all the data we had to mobilize to understand this issue of the interaction between water mobility and human mobility. It explains what data were needed to reconstruct daily mobility at the scale of a department, understand the "floodability" of road networks, analyze the effects of individual decisions on drivers' exposure in the event of flooding, and evaluate the impact of extreme events on residential choices, land use, and regulations. It also explains what our approach was to integrate all these data, in particular by mobilizing the notion of scale.

The structure of the book, therefore, allows for an *à la carte* reading, either linear or by drawing from the different chapters offered according to needs and desires.

Once again, we would like to thank all the people and structures who made it possible to undertake these works, the writing of this book and its publication. First of all, our thanks go to the ANR, and more specifically to the *programme Société et Environnement* (Society and Environment Program), which, through its decision n° ANR-12-SENV-002, financially supported the *MobiClimEx* project that provided the framework for the writing of this book.

We also thank the ISTE for having given us their trust and support in this long-term exercise, despite the difficulties and delays that the writing of this second volume has encountered.

We would like to thank the authors who, although the *MobiClimEx* project has been completed, have maintained their efforts to make it possible to write this second volume.

We hope that this book will share with its readers all the richness and intellectual pleasure we have had in its writing.

References

[AUB 14] AUBLET B., L'action en situation d'urgence: facteurs d'efficacité dans la gestion du réseau routier en cas de crues rapides. L'exemple du département du Gard, PhD thesis, Université Joseph-Fourier-Grenoble I, Grenoble, 2014.

[CRE 09] CREUTIN J.D., BORGA M., LUTOFF C. *et al.*, "Catchment dynamics and social response during flash floods: the potential of radar rainfall monitoring for warning procedures", *Meteorological Applications*, vol. 16, pp. 115–125, 2009.

[CRE 10] CRETON-CAZANAVE L., Penser l'alerte par les distances. Entre planification et émancipation, l'exemple du processus d'alerte aux crues rapides sur le bassin versant du Vidourle, PhD thesis, Université Joseph-Fourier-Grenoble I, Grenoble, 2010.

[CRE 13] CRETON-CAZANAVE L., LUTOFF C., "Stakeholders' issues for action during the warning process and the interpretation of forecasts' uncertainties", *Natural Hazards and Earth System Sciences*, vol. 13, no. 6, pp. 1469–1479, 2013.

[DEB 16] DEBIONNE S., RUIN I., SHABOU S. *et al.*, "Assessment of commuters' daily exposure to flash flooding over the roads of the Gard region, France", *Journal of Hydrology*, vol. 541, pp. 636–648, 2016.

[LUT 18] LUTOFF C., DURAND S., *Mobility in the Face of Extreme Hydrometeorological Events 1*, ISTE Press Ltd, London and Elsevier Ltd, Oxford, 2018.

[NAU 12] NAULIN J.P., Modélisation hydrologique distribuée pour la prévision des coupures de routes par inondation: application au département du Gard, PhD thesis, École centrale de Nantes, Nantes, 2012.

[RUI 07a] RUIN I., Conduite à contre-courant. Les pratiques de mobilité dans le Gard: facteur de vulnérabilité aux crues rapides, PhD dissertation, Université Joseph-Fourier-Grenoble I, Grenoble, 2007.

[RUI 07b] RUIN I., GAILLARD J.-C., LUTOFF C., "How to get there? Assessing motorists' flash flood risk perception on daily itineraries", *Environmental Hazards*, vol. 7, no. 3, pp. 235–244, 2007.

[RUI 08] RUIN I., CREUTIN J.D., ANQUETIN S. *et al.*, "Human exposure to flash-floods – relation between flood parameters and human vulnerability during a storm of September 2002 in Southern France", *Journal of Hydrology*, vol. 361, no. 1–2, pp. 199–213, 2008.

[RUI 12] RUIN I., LUTOFF C., CRETON-CAZANAVE L. *et al.*, "Toward a space-time framework for integrated water and society studies", *American Meteorological Society*, vol. 93, no. 10, pp. 89–91, 2012.

[RUI 14] RUIN I., LUTOFF C., BOUDEVILLAIN B. *et al.*, "Social and hydrological responses to extreme precipitations: an interdisciplinary strategy for post-flood investigation", *Weather, Climate and Society*, vol. 6, no. 1, pp. 135–153, 2014.

[SHA 16] SHABOU M.S., Extrêmes hydro-météorologiques et exposition sur les routes: Contribution à MobRISK: modèle de simulation de l'exposition des mobilités quotidiennes aux crues rapides, PhD thesis, Université Grenoble Alpes, Grenoble, 2016.

[SHA 17] SHABOU S., RUIN I., LUTOFF C. *et al.*, "MobRISK: a model for assessing the exposure of road users to flash flood events", *Natural Hazards and Earth System Sciences*, vol. 17, pp. 1631–1651, 2017.

1

Interdisciplinary Travel

1.1. Introduction

The constant change that characterizes the 21st Century society, subject to unprecedented changes in its daily environment, is prompting science and the production of knowledge to a profound renewal [PET 08]. While the 20th Century was marked by significant advances in knowledge in each discipline, the challenges that emerged in the last decades of this century were pushing researchers to move beyond their discipline in order to better understand the changes taking place and to support society in these changes. Interdisciplinarity thus became imperative in scientific research, as the introduction of the special issue of *Nature* in 2015 on the subject claimed: "To solve the grand challenges facing society – energy, water, climate, food, health – scientists and social scientists must work together" [NAT 15].

But how can we create an interdisciplinary approach that works? What does this imply in terms of skills, as a particular posture? How does this transform the dominant scientific approach to date?

To provide some answers to these questions, we propose, in this chapter, a journey in interdisciplinarity. Like explorers, we propose to start from the singular experiments that we have carried out as part of a team, grouped together over several consecutive projects and seeking to understand the socio-hydro-meteorological dynamics that, during extreme events, can lead to disasters or on the contrary avoid them. We will complete and enrich this feedback with a literature review on the issue of interdisciplinarity. In terms

Chapter written by Céline LUTOFF, Jean-Dominique CREUTIN, Séverine DURAND, Isabelle RUIN, Sandrine ANQUETIN and Brice BOUDEVILLAIN.

of experiments, we will rely particularly on the MobiClimEx project on social mobility in the face of extreme hydrometeorological conditions. This project, funded by the ANR, took place between 2013 and 2017 and brought together researchers in hydrology, meteorology, geography, sociology and legal sciences. Beyond the expected difficulties, we would like to focus our attention on what has worked well in this interdisciplinary work and what has really allowed the researchers involved to move beyond their specific logic to discover together other ways of thinking.

This chapter thus proposes to specify the paths to be taken, the necessary postures and the paradigm shifts that take place in the interdisciplinary journey. In the first step, we propose to clarify what interdisciplinarity brings in relation to other approaches and why it is important today. We will then look at the obstacles and stumbling blocks that interdisciplinary travelers may have encountered on their journey. We will then explain what we think is essential to undertake in such a journey and how to avoid these obstacles: the choice of destination, the qualities expected of the traveler and the content of their luggage. We will then conclude on the essential need to prepare such a trip and propose some tools to produce successful interdisciplinarity.

1.2. Why this interdisciplinary journey?

The Global Research Council (GRC)[1] chose in 2015 to make interdisciplinarity one of its two annual working themes in order to define a common position on interdisciplinary research or IDR[2] [RYL 15]. The latter is defined as research that combines two or more disciplines to produce a common body of research [GLE 16].

The history of science has been marked since antiquity by a structuring of the production of knowledge in highly differentiated disciplines, which has continued to grow stronger over time [LED 15]. While the use of the term interdisciplinarity appeared in the 1940s (notably in 1937 in a sociology article), it was not until the 1970s that an academic movement was clearly organized in interdisciplinarity around issues such as environmental

1 The GRC is a federation of more than 50 national research funding agencies, with representatives from many countries including Brazil, China, Japan, Russia, the United Kingdom and the United States. The GRC is also part of the Science Europe consortium, an association of research funding agencies in Europe [RYL 15].

2 IDR for Inter Disciplinary Research.

protection in particular [LED 15]. The movement began to spread in Europe, as evidenced by the publication in 1994 of the book by [GIB 94] which predicted a new form of knowledge production and an increase in interdisciplinary research, as the only way to answer the complex questions facing society. This book then paved the way to a new method of producing knowledge and doing science.

In the special issue of *Nature* on interdisciplinarity, [RYL 15] identifies three reasons for following this path. First, as already mentioned, the current problems are complex. This is the case for climate change and food or water security, which cannot be properly addressed by a single discipline. Second, according to [LAK 07], the major current advances are rather at the boundaries of disciplines. Indeed, when evaluating research projects that have an impact outside the academic field, 80% are interdisciplinary research results [RYL 15]. Third, interdisciplinary work also benefits each of the disciplines, broadening their own boundaries and horizons.

While the interdisciplinary approach appears to be crucial to address the problems that society is currently facing, how far can it go? In [BAR 17], a literature review conducted over the period 1990–2014, it has been noted that interdisciplinarity can take many forms. They then observe that individual interdisciplinarity (multiple skills acquired by a single researcher) is more developed than the structuring of interdisciplinary teams involving several researchers. They also show that while interdisciplinarity is generally progressing, work between very distant disciplinary fields remains underdeveloped. However, current environmental issues, concerning energy, water, climate, food, and health, require this type of approach. This form of interdisciplinarity is sometimes called “radical interdisciplinarity” [EVA 06], [MAR 08], or “distal interdisciplinarity” [YEG 15], and combines very distant disciplinary fields such as earth sciences, life sciences, engineering sciences, and human and social sciences. However, the effort required to construct this radical interdisciplinarity may appear very complicated and time-consuming, which may discourage some researchers. This is what [KLE 10] and [RYL 15] report: *Interdisciplinary research that involves neighbour disciplines is much more common, and significantly easier to develop, than areas in which the disciplinary stretch is vast and the logistics and intellectual challenge more demanding* [RYL 15, p. 315].

However, this is the challenge we have been taking up for more than 15 years at the Université Grenoble Alpes, combining environmental sciences (hydrology, climatology), engineering sciences (computer science), and human and social sciences (geography, sociology, and economics), as do other teams around the world [LED 15]. This effort is voluntary because we are all convinced that the association of several researchers, exploring together a common problem and each bringing a particular perspective, makes it possible to construct new knowledge, which goes far beyond that which could be produced by each person alone. Here, as in system analysis, the whole is more than the sum of its parts. Several authors agree that radical interdisciplinarity is a necessity, an essential evolution of science so that it can answer the major current questions facing our societies [BAR 17], [EVA 06], [NAT 15], [PET 08].

But beyond this academic necessity, the challenges of the 21st Century also require us to cross other boundaries, hitherto more or less hermetic. This is particularly the case for the boundary between scientific knowledge and practical knowledge, relating to political decision-making and concrete action on the ground [HAD 08]. Crossing this boundary is transdisciplinary research, which involves not only opening up to different knowledge within the academic sphere (interdisciplinarity) but also the integration of non-academic knowledge and has its own characteristics [HAD 08]. This is in line with the ongoing reflection on "knowledge in Mode 2" proposed by [NOW 06] in particular, which advocates the construction of knowledge that integrates the contribution of the various academic disciplines but also that of politicians, professionals in the field and society as a whole [PET 08]. To summarize these different approaches and better understand the levels of integration, we would say that multidisciplinarity (or pluridisciplinarity) assembles disciplines, each one of those continuing to mobilize independently its own concept, method and tools. Meanwhile, interdisciplinarity assembles disciplines that might be scientifically distinct, but with the common goal to produce combined and new approaches, tools or methods to solve complex problems. Sometimes these problems are so complex that academic knowledge alone does not allow us to understand all their dimensions. In this case, the transdisciplinary approach involves both teams of researchers and operational actors in an attempt to better understand how to approach them.

This new approach to research, through the integration of different disciplines and professional actors, has been gradually developing since the beginning of the 21st Century. We are witnessing the emergence of a new mode of research organization, called "challenge-led (or 'problem-led') research organization" [GIB 19]. Funding agencies in several countries are placing some of their programs in this perspective. This is the case, in particular, for the Research Council in Great Britain (Research Councils UK

2018). In return, academic and research institutions organize themselves by proposing major projects in which interdisciplinarity is highlighted as a necessity to meet these major societal challenges. This has led to major projects at the world's leading universities, such as University College London's Grand Challenges, Warwick University's Global Research Priorities, Delft University of Technology's Research Initiatives Program, and Princeton's Grand Challenges.

On a French scale, the academic system was strongly influenced by Napoleon, who organized the public service around the *corps d'Etat*, the state body, thus linking the academy to the field with the help of large institutions of which IRSTEA and IFSTTAR, for example, are descendants. These establishments have in their own way been involved in the "challenge led" organization since the beginning of the 19th Century [PIC 14], [SHI 78]. More recently, the IDEX projects (excellence initiatives) of the *Programmes d'investissement d'avenir*, PIA (Future Investment Programs)[3], deployed following the 2010 Finance Act in the major university sites selected, are placed in the same perspective and seek to develop interdisciplinarity and transdisciplinarity to meet these major societal challenges. We are, therefore, witnessing a significant change in the production of knowledge, with a focus on contributing to the definition of "socially robust" solutions to major problems or challenges [NOW 03]. This new vision of research thus makes interdisciplinarity and transdisciplinarity essential approaches. Nevertheless, several obstacles still exist, both from the point of view of researchers and institutions, to the development of these approaches.

1.3. Barriers to interdisciplinary and other "travel" difficulties

The special issue of the journal *Nature* in 2015 drew attention to the fact that interdisciplinarity is difficult to achieve, fund, evaluate and publish: *it is hard to deny that the scientific system – from funding streams and academic rewards to university departments and journals – does not encourage much overlap between disparate subjects* [NAT 15]. While they encourage governments, research funding bodies, and academic institutions to make the "no-man's-land of interdisciplinarity" a more welcoming place for the career development of researchers, they point out that there are many obstacles.

3 For more information, see www.enseignementsup-recherche.gouv.fr.

The first of these is that interdisciplinary projects take longer than others and therefore cost more, which is why this *Nature* publication said: *True interdisciplinary science cannot be rushed* [NAT 15]. Indeed, for a researcher, working in interdisciplinarity implies a first incompressible phase of inter-knowledge and acculturation to other disciplines. This phase, which is generally non-productive in terms of scientific productivity indicators, may take more or less time, depending on the distance between the disciplines involved. A "proximity" interdisciplinarity within the Earth sciences or social sciences will potentially take less time than a radical interdisciplinary project. Indeed, the tools and methods generally remain comparable in the first case. The human and social sciences, for example, mobilize an arsenal of methods and tools that can be used in geography, environmental psychology and sociology: survey practices, in particular, are part of these multifunctional tools. While the use and modes of analysis differ, acculturation between these disciplines is thus facilitated by the mobilization of these common tools. On the other hand, bringing together researchers in hydrology and sociology requires a longer phase of inter-knowledge. Indeed, it is a question of associating researchers with different scientific cultures, mobilizing methods, tools, and concepts unknown to others. Regular communication between researchers, clarification of the issues at stake, but also of the vocabulary used, is essential to ensure the coherence of a radical interdisciplinary project.

A second obstacle, particularly concerning radical interdisciplinarity, stems from the difference in scientific culture between disciplines. The very way in which what can be considered a "problem" is referred to differs from one discipline to another [PET 08, p. 598]. Similarly, the methods for investigating these problems are different and may sometimes be opposed. What can the qualitative approach bring when we are convinced that only quantitative methods allow modeling, which has become essential? Or conversely, what is the point of a physics-based approach when we are convinced that most of the current problems are eminently social?

In this perspective, one of the most frequently cited problems in terms of interdisciplinarity concerns the role of the human and social sciences (HSSs) in interdisciplinary construction. As several authors point out, HSS researchers regularly feel "instrumented" (or instrumentalized), confined to playing a facilitator role for the deployment of technical solutions defined by the "harder" disciplines by removing the cultural, psychological, and/or social barriers that these solutions may encounter in their implementation [ASP 08], [PET 08]. Their role in the interdisciplinary team can sometimes be reduced to providing a single ethical endorsement to the project, denying the scientific questions specific to social science researchers [ASP 08]. This is particularly the case when HSS research is limited to a single researcher within the interdisciplinary team. Nevertheless, even when an entire team of researchers is mobilized, they are encouraged to adopt methodologies or approaches that provide results that can be understood by the natural, life sciences, or technical sciences and integrated with their results. Quantitative methods are thus more easily integrated into an interdisciplinary approach, particularly in the modeling that has become massively popular, favoring certain disciplines (such as economics in particular) or certain disciplinary trends and leaving aside more qualitative methods [PET 08]. These cultural differences imply that the interdisciplinary approach requires a phase of negotiation of the target issues and the methodological framework to be mobilized to address them [GLE 16], [PET 08].

A third obstacle noted by [PET 08] is that from an academic point of view, the silo vision of disciplines is sometimes accompanied by the belief that science must necessarily be done without any practical utility [PET 08, p. 597], [SNO 64]. Interdisciplinarity, driven by the need to respond to contemporary societal problems, therefore falls into the category of applied research and is therefore relayed to the lowest level in the academic hierarchy. This is forgetting that interdisciplinarity is also capable of theoretical and conceptual developments that can structure disciplinary knowledge itself [PET 08]. Nevertheless, the system of academic evaluation and career development of researchers still remains very focused on a quantitative approach to disciplinary results and publications, which does nothing to promote the development of interdisciplinarity [GIB 19], [RYL 15]. In some cases, even valuing interdisciplinary work within a discipline is a challenge, as it is considered too limited in depth in the discipline in question. And indeed, the necessary simplification of the contributions of each discipline to interdisciplinary construction[4] can devalue these works in the eyes of experts in a particular field. This can block the entry into a research career for doctoral or post-doctoral students involved in interdisciplinarity.

The conditions for publishing the work thus constitute a fourth important obstacle to interdisciplinarity. First, the evaluation of interdisciplinary work is a challenge for publishers: how to apply the principles of peer review and

4 On this necessary disciplinary simplification for interdisciplinarity, see in particular Chapter 7 of this volume.

identify, outside disciplinary channels, specialists capable of evaluating interdisciplinary work [RYL 15]? To address this difficulty, some multidisciplinary journals have emerged on specific themes and have succeeded in mobilizing specific communities. However, these journals have long had a lower impact factor than mono-disciplinary journals; similarly, interdisciplinary articles are less cited than those in a single discipline [BAR 17]. Interdisciplinary research is considered to be less innovative and too little advanced in terms of theory, which by definition must be negotiated between the different disciplines. Contrary to recent studies [RYL 15], interdisciplinary research is thus sometimes considered to have a lower capacity for innovation than disciplinary research. This is particularly the case when the assessment of these innovative capacities is based solely on the place given to interdisciplinary publications in the scientific literature [PET 08]. Consequently, [RYL 15] thus highlights the very great difficulty in assessing interdisciplinary excellence; this evaluation is generally based on a disciplinary approach: *Items submitted to discipline-based REF (Research Excellence Framework) panels under-represented the quantity of top interdisciplinary research published by UK researchers in some fields. (...) This suggests that researchers perceive inter-disciplinary research to be vulnerable to discipline-based assessment.* This difficulty in assessing excellence also severely penalizes researchers working in interdisciplinarity, whether for obtaining a position or career development in French research institutions, where evaluations are carried out by discipline (e.g. at the level of the French National University Council to obtain a teaching qualification or an authorization to conduct research).

However, since the early 2000s, there has been an increase in the number of interdisciplinary environmental journals, which have gradually taken a prominent place in all scientific journals, with impact factors now competing with some disciplinary journals [PET 08].

Similarly, the research funding system, the fifth obstacle to interdisciplinarity, is still very discipline-based. This leads to the following observation by [RYL 15]: *it [is] significantly more difficult to gain funding for [interdisciplinarity] than for mainstream activity.* The history of the development of science indeed gives pride of place to competition, rather than collaboration, and interdisciplinarity appears to be at odds with quality scientific development and raises fears: *Some fear that it drains funds, time, and energy from "core" disciplines* [RYL 15]. The cost of interdisciplinary construction is another financial barrier. Interdisciplinary work requires regular and intense collaboration between disciplines, involving the creation of space-time for exchanges and discussions that can be permanently mobilized throughout the duration of the interdisciplinary construction. This results in a reorganization of the research work as a whole, without being sure of the gains that each of the disciplines will be able to make. This cost of entering interdisciplinarity can then be seen as risky and discouraging for both researchers and institutions [PET 08].

Thus, at the end of this work by the GRC in 2015, the authors acknowledge that many research funding agencies have not necessarily defined a specific policy to support interdisciplinarity by that date, although some had sought to promote "good practices" in this area [GLE 16]. Beyond the recommendations to provide the time and resources needed to create interdisciplinarity, the authors of the GCR's final report encourage funding agencies to define specific evaluation modalities for interdisciplinarity, in order to move beyond the current logic of disciplinary evaluation and create appropriate indicators and evaluation tools [GLE 16]. However, it will take time to implement these profound changes in the way knowledge is produced and evaluated; three years after this report, it is still difficult to see any real implementation of these recommendations.

Despite all these real obstacles to interdisciplinarity, more and more researchers, particularly young researchers, are interested in the approach and do not hesitate to commit themselves to it, with the essential driving force of the desire to tackle the complex problems of our societies and to provide new knowledge to help solve them. Also, in the following section, we propose to provide some background information and some recommendations for engaging in the interdisciplinary approach, as we have experienced it on the issue of mobility in the face of hydrometeorological extremes.

1.4. Beginning the journey

Based on their interdisciplinary experience on water management issues in Australia, [BRO 15] identify five principles of successful interdisciplinarity: 1) defining a common mission; 2) training interdisciplinary researchers (T-shaped researchers); 3) fostering constructive dialogue; 4) having institutional support; and 5) linking research, practice, and policy-making. Our own experience of interdisciplinarity leads us to join these principles as we will see, but to formulate them a little differently. To continue the metaphor of travel, we will first discuss the choice of destination, the specific skills of the traveler, and the methods and tools that they must take with them on this adventure.

1.4.1. *Choosing a destination*

Choosing a destination here corresponds to the first principle of [BRO 15]: to define a common mission. This requires specifying the purpose of the research, which must be defined broadly enough to allow the different disciplines to project themselves into it and present a level of complexity that requires the collaboration of these different disciplines.

In our case, this common object is the risk of flash flooding. We have chosen to approach this subject in all its complexity, defining it as an interaction between water dynamics and human dynamics at different spatial and temporal scales (see Chapter 8 of Volume 1). Our common mission is to understand how this interaction can sometimes lead to human disasters or, on the contrary, to an event with no consequences, despite its intensity. Our objective is to identify the hydrological, meteorological, human, legal, and institutional ingredients that lead to the first situation rather than the second.

This definition of a destination is constructed from the beginning of the project, of course, but it is defined throughout the journey. Permanent and constructive exchanges between the passengers involved (principle no. 3 [BRO 15]) are a necessary condition to both stay the course and clarify our objectives. These exchanges can be organized within the framework of the project through meetings fixed according to a predefined rhythm (monthly, semi-annual, and annual). However, these meeting points, while important for the coherence of the project, are not sufficient for a real co-construction of the common objectives. Despite technological progress in terms of communication and the possibility of easy remote exchange, co-presence is an essential condition for this mutual construction of the outlines of the project's final destination. These outlines are refined during seminars and planned meetings, but they are also refined on a daily basis, sometimes in informal exchanges (over a cup of coffee, for example) according to the current readings or analyses in progress. In our experience, we clearly perceived the difference in ownership of the project's missions and objectives between team members gathered in the same place and working together on a daily basis and other members, more distant geographically.

However, co-presence is not everything and travelers must also have certain characteristics if the final objective is to be truly achieved.

1.4.2. *The interdisciplinary traveler*

Among the qualities expected of the interdisciplinary traveler are of course openness to other disciplines and curiosity. However, as [PE 10] indicates, this curiosity can sometimes mask a more or less conscious desire to bring something to the other discipline, more than to take from it. In a kind of superiority of the discipline, this attitude may limit interdisciplinarity to unilateral exchanges likely to jeopardize the experience. To avoid this pitfall, the traveler of the interdisciplinary must demonstrate a desire to understand other scientific cultures and to integrate certain approaches or postures. Let yourself be imbued with the scientific culture of others to enrich your own; build new knowledge, rich in the contributions of the different paths explored; let new ideas emerge by sharing data that are *a priori* heterogeneous, but which all contribute to a better understanding of the problem at hand[5]. It also implies identifying one's fears toward other disciplines and appeasing them: accepting not to master everything, but not giving up on understanding the general approach, the main concepts.

The interdisciplinary traveler may initially come from any discipline of origin and be interested in the major societal issues and challenges posed by the interactions between the environment and society. Let us take the concrete example of understanding the risk of flash flooding that we discuss in this book. It is commonly accepted that this risk is the conjunction between a hazard (the natural phenomenon) and exposed elements (populations and assets) characterized by a specific vulnerability (or on the contrary resilience capacities). These three components of risk (risk, exposure, and vulnerability) are now used to define climate change as a risk which the IPCC reports [GIE 14]. In our case, risk analysis is a matter of concern for geosciences and more specifically climatology, meteorology, and hydrology. Interdisciplinary work is already required at this level if the risk is to be understood as the result of hydrometeorological coupling in a changing climate context. How do complex atmospheric processes lead to intense rainfall events that generate flash floods? How can these rains cause rivers to rise, depending on the state of the watershed? To complete this approach, an exposure analysis should be undertaken: how many people, buildings, roads, supply networks, and so on

5 On this point, see Chapter 8 of Volume 2 (Process).

are exposed to the phenomenon? For individuals, exposure assessment is often limited to an approximation based on existing data on the counting of the population at their place of residence (usually INSEE data). However, for a more detailed assessment, the analysis must take into account the movements of these populations over time and space. As shown in Chapters 4 and 5 of this book, this analysis is not trivial and also requires the combination of the skills of demographers, geographers, and environmental psychologists in particular. Finally, the analysis of the vulnerability of these exposed elements (or their resilience) requires not only engineering approaches when considering technical buildings and networks but also economic, sociological, ethnographic, geographical, and psychological approaches to address the vulnerabilities of populations and their activities. It can, therefore, be seen that taking into account all the components of risk requires the contribution of several disciplines.

Nevertheless, to really understand what is at stake at the time of an event, the mere juxtaposition of the contributions of each of these disciplines is not enough! To trace the thread of what happens during a major hydrometeorological event, it is necessary to understand the interactions between the mobility of people (who are the people at the site of the event at the time of its occurrence) and the mobility of water (what are the characteristics in terms of flow velocity, water levels, and so on at each of the points in the watersheds affected by the episode). Multidisciplinarity alone is not enough in this case, and it is necessary to provide the methodological, even theoretical, means to identify the socio-hydro-meteorological processes that lead to these interactions. In the following section, we propose to detail the tools we have used in our case to build this interdisciplinarity.

1.4.3. *The traveler's backpack in interdisciplinarity*

On the basis of our experience, we propose here to explore not only the tools mobilized, but also the useful postures to enter into a real radical interdisciplinarity.

1.4.4. *The toolbox*

Once the researcher is convinced of the value of working in interdisciplinarity, one of the first questions is: how do we do it? What are the methods and tools that really allow us to go beyond the mere juxtaposition of the contributions of each of the disciplines to create new knowledge? First, as already mentioned, it is advisable to define a privileged space–time to share knowledge and negotiate the concepts and methods to be promoted within the framework of the interdisciplinary project. This requires numerous exchanges whose objective is to understand each other, to agree on the terms to be used, and their definition, but also on the methodological tools to be used.

Concerning our analysis of the risk of flash flooding, these reflections on the toolbox were carried out within the framework of a working group, set up in 2010 at the Université Grenoble Alpes and bringing together researchers from several laboratories in geosciences (hydrologists, seismologists, etc.) and human and social sciences (geographers, economists, political scientists, sociologists, etc.). This is the AMETIhST[6] seminar, which meets every month with the objective of sharing theoretical and methodological tools, and testing their relevance and manageability in an interdisciplinary manner. This involves reading together articles chosen collectively as a possible basis for our interdisciplinary work on a selected theme and generally renewed each year. The rule for joining AMETIhST during the year is, therefore, to have read the articles discussed. The idea is that everyone takes ownership of the shared references and takes their own disciplinary viewpoint: these points of view are then exchanged, each identifying the contributions and limits of the methodological proposal on the theme concerned. Once everyone has taken ownership of this proposal from a theoretical point of view, we test its usefulness as an interdisciplinary tool to address the chosen theme and move toward the production of new knowledge.

To give some more concrete examples, we have thus chosen to analyze the evolution of pastoralism in West Africa under climate constraint. The conclusive test of the mobilization of this spatio-temporal approach to address issues combining climate problems and social evolutions subsequently made it possible to retain the spatio-temporal graph as a basic tool for the analysis of interactions between social dynamics and hydrometeorological dynamics. More recently, we have chosen to question the notion of scale in the interactions between social mobility, short-term

6 AMETIhST for *Anticipation, Mobilité, Echelles spatio-Temporelles, Intégration, hydrométéo, Société, Territoires* (Anticipation, Mobility, Spatial-temporal Scales, Integration, Hydrometeorology, Society, Territories). For more information, visit: http://www.ametihst.fr.

(daily mobility) and long-term (residential mobility) views, and extreme hydrometeorological phenomena[7].

The experience gained through AMETIhST has made us aware of certain realities of interdisciplinary work: it is only possible if: 1) a certain level of *impregnation* by other disciplines is accepted; 2) capacities are developed to *transpose* certain concepts or methods from one discipline to another; and 3) simplifications and other adjustments are tolerated which allow the *joint production* of knowledge with other disciplines. We now propose to review these conditions by giving some concrete examples of their implementation from our experiences.

1.4.5. *Impregnation*

To accept impregnation is to tolerate venturing into another scientific universe and to take the risk of losing one's points of reference in order to find others. As we have already mentioned, this starts with vocabulary issues: contrary to what one might think, talking about a common research object with a colleague from another discipline can be very troubling, simply because the terms are not used in the same way. Where a hydrologist or meteorologist sees exposure as one of the components of the risk equation, a sociologist or geographer will see it as the result of a whole system of factors, whether economic, social, cultural, or organizational, which cannot be easily understood by a single synthetic value (a cost or a number of people affected). Understanding each other then implies that the former accepts that exposure cannot be summarized in a single number and the latter makes the effort to summarize all the factors in order to be able to integrate them into the risk equation.

However, to achieve this, impregnation also means not hesitating to ask "naive questions": every discipline has its self-evident facts that it no longer seems necessary to re-explain when we are between specialists. But these facts are not necessarily true for researchers in other disciplines. It is sometimes necessary to go back to basic explanations to allow everyone to fully understand what the meaning behind the words is. Using the term "watershed" with a sociologist implies explaining to him what this term refers to in hydrology. Similarly, when a geographer speaks of representation or perception, it is often necessary for him to clearly define

7 Some of the results of this work on scales are presented in Volume 1 of this book series.

what each of these terms means, both for other researchers in the human and social sciences and for researchers in other disciplines. However, this exercise requires some practice to identify which concepts are not self-evident. That's why asking "stupid" questions is important! Even when a term seems common and quite clear to everyone, having its definition clarified, its use in each discipline often reveals very interesting surprises and discussions! An external and "naive" look can sometimes reveal certain misunderstandings within a discipline itself.

1.4.6. *Transposition*

Impregnation is accompanied or may be followed by a transposition phase, i.e. the appropriation and integration of certain concepts or tools from another discipline into its own approach to the subject under study. In our case study, this transfer led to questions about the notion of a watershed. *A priori*, for a social science researcher, this notion does not necessarily appear to be very useful and can be identified from the outset as a specificity of the hydrological approach. However, we have examined whether this concept could be useful in analyzing the circumstances of death of individuals in the event of flash flooding [RUI 07]. This involved applying the spatio-temporal analysis by watershed, used by hydrologists to characterize the flash flooding domain[8], to identify deaths recorded during the 2002 flood in the Gard.

As shown in the following figure, which is based on [RUI 09] and completed, this approach has made it possible to highlight certain characteristics of social vulnerability to flash flood phenomena. A first category of vulnerable population, well identified in the literature on natural hazards, and which concerns more particularly the elderly, surprised by the flooding in their homes, is identified. The latter are more concerned in the case of 2002 by the reaction of relatively large watersheds (>1,000 km^2), with response times of around 6 hours and more (>400 minutes). However, the analysis by watershed made it possible to identify a second category of population, less often identified as vulnerable: it is a rather young population, 43 years-of-age, middle age, mainly men, who lost their lives outdoors, while traveling on foot or by car. The analysis by watershed also identified that this category of vulnerability is sensitive to very small basins (<20 km^2 here), with response times in the order of a few minutes (10 minutes–1 hour). We are here in the most critical situations of flash

8 See Lutoff and Durand 2018, Volume 1, p. 73.

floods, on watersheds that are often dry in these regions of south-eastern France, and which can react violently to very localized storms.

The vulnerability of this category of population is not specific to France and can be found in some references [ANT 01], [BOU 06], [GRU 77], [GRU 00a], [GRU 00b], [STA 94]. Nevertheless, the use of the watershed approach, inspired by work with fellow hydrologists, has made it possible to highlight the specific circumstances of these deaths in very small watersheds, characterized by very short response times, for which warning is very difficult to achieve. These results led us to ask our hydrologist partners to find out more about the dynamics of these watersheds. We then realized that the use of hydrological knowledge available at these scales remained very limited, partly because of the difficulty of instrumenting the multitude of small watersheds, but also because it was thought that these scales were not relevant to the flood issue. Since then, significant work has been carried out, also in hydrology, to supplement this knowledge, by mobilizing tools and methods different from those previously used (see, for example, [MAR 10], [MEN 12], [LOO 12]).

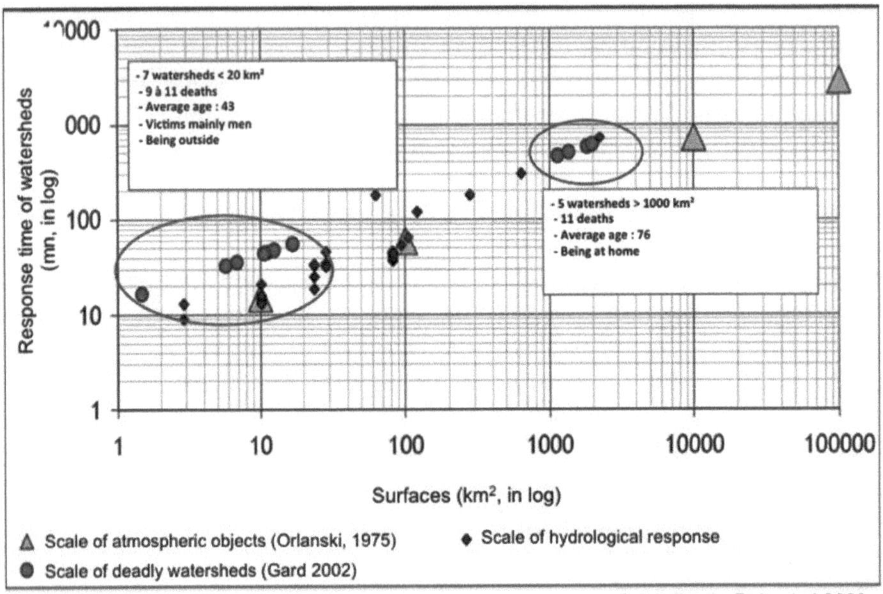

Figure 1.1. *Circumstances of deaths following the 2002 flood in the Gard and vulnerability of the populations. For a color version of this figure, see www.iste.co.uk/lutoff/hydrometeo2.zip*

The transposition of the watershed approach, on which social science researchers would not have chosen spontaneously to analyze the circumstances of death, has therefore made it possible to produce new knowledge in this field, which is now recognized elsewhere (see in particular [BOI 13], [SCO 12], [VIN 11]), but also in hydrology. This round trip of questioning and mobilizing concepts leads us to specify the conditions for jointly producing knowledge developed in interdisciplinarity.

Joint production

Beyond this example of transposition, the interdisciplinary experience we have had for more than 10 years has enabled us to jointly produce a reference framework for analyzing social mobility in the face of flash flooding in an interdisciplinary way. In this respect, the discussions carried out within the framework of the AMETIhST seminar were particularly beneficial. Indeed, we have taken advantage of methodological tools such as the space–time scheme proposed by [HOL 01] or the scale approach of [CAS 06] supplemented by [GRO 07] to understand our common object of study. This has allowed us to create new approaches, such as the calculation of anticipation time [LUT 16]. The idea is to assess whether the actions taken by individuals exposed to flash flooding are adapted to the specific timing of the event, which we have just seen is highly dependent on the size of the watersheds considered.

The analysis of the actions taken by individuals at the time of the flood required the development of an original method, inspired by a demographer's tool: individual trajectories, which consist of *retrieving and ordering events that occurred to the same individual in the observation area* [BRI 05]. We sought to apply this method to the individuals concerned by an event, i.e. on the precise area and in the limited time of the event. A data collection tool was developed and tested in several feedback cases, following flood events between 2010 and 2014 (for more details, see [RUI 14]). The data collected were then processed by mobilizing actograms – a tool used by geographers specializing in mobility to better understand the characteristic times from a social point of view (starting times at work, lunch, sleep, etc.). Adapted to our problems, these actograms have enabled us to identify certain characteristic phases in the actions undertaken in response to the flood by different individuals in the same place: to be informed, to organize themselves to cope with the event (prevent their loved ones, protect certain assets, etc.), or to protect themselves.

We also mobilized these categories of actions and examined the precise timing at which they were implemented in each of the watersheds considered. This is what led us to calculate anticipation times. Indeed, as we have shown, the time available to undertake the different categories of action is not the same depending on whether one is located in a small or large catchment area. We, therefore, sought to determine how the succession of activities was undertaken according to the size of the watershed under consideration, and therefore according to the time available to implement them. As shown in Figure 1.2, taken from [LUT 16] and completed, all categories of actions, represented by symbols (crosses, circles, or triangles) are undertaken regardless of the size of the watershed. In some cases, actions were even taken several hours before the peak flood level, the reference point for our calculation, characterized by the line of abscissa in the figure[9]. It can be seen that these anticipated actions concern both small and large watersheds. This raises questions about what led people to inform, organize, or protect themselves well in advance of the event, especially when the reaction time to the watershed is very short.

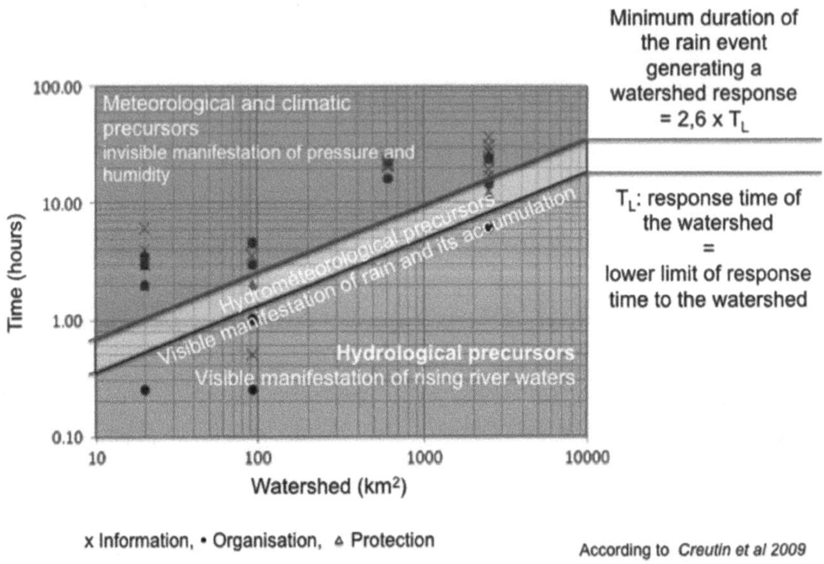

Figure 1.2. *Anticipation time and notion of precursors – actions are represented by symbols (cross/information, rounds/organization, or triangles/protection). For a color version of this figure, see www.iste.co.uk/lutoff/hydrometeo2.zip*

9 For more details on this calculation of the anticipation time and on the use of the flood peak as a temporal reference point, see [LUT 16].

We then mobilized the knowledge of our hydrometeorological colleagues to propose a characterization of the identifiable precursor signs in the environment, according to the phases of the flood process up to its peak danger (the peak flood). Chapter 8 of this volume gives all the details on this notion of physical precursors for the analysis of social action. The spatio-temporal diagram shown in Figure 1.2 enables here a representation of what precursors can be by dividing them into three categories corresponding to three specific phases of the flood. "Meteorological and climate" precursors are the earliest and not the easiest to identify in the environment. As invisible manifestations of pressure and humidity fields, meteorologists here use prediction models that make it possible to identify, sometimes several hours in advance, weather conditions favorable to intense rainfall that generates flash flooding, based on atmospheric observations of pressure, temperature, and humidity. Nevertheless, even without these elaborate technical means, some residents manage to undertake certain actions (often with regard to information or organization) on the basis of signs or impressions that they can perceive. Some amateur meteorologists[10] are able to identify certain signals in the environment that, even if they are weak, lead them to a vigilant posture. Then come the "hydrometeorological" precursors or, to put it simply, the heavy rain that will provoke the response of the watershed on which the population is located. The duration of precipitation required to provoke the response of the different watersheds varies, of course, according to their size. Finally, hydrological precursors appear: this is the concrete manifestation of the watershed's response, which is reflected first in the rise of water in the riverbed, then gradually in the process of overflowing and flooding the surrounding land. These overflows can occur well upstream of the peak flood, depending on its intensity. For exposed individuals, this translates into a vision of this growing river that can more or less quickly invade their road or building.

In this set of balances between the disciplines that we have observed in interdisciplinary work, this last observation leads us in the social sciences to look for a way to collect even more data on social responses in crisis situations, in order to understand how these precursors are (or are not) perceived and what attitudes they cause in exposed people. This work would then make it possible to identify the levers of preventive actions and to integrate these perceptual dimensions into preventive information campaigns, for example.

---

10 In the noble sense of the word.

1.5. Conclusion: a trip to prepare yourself for!

At the end of this journey in interdisciplinarity, we would like to return to our initial questions: how do we create an interdisciplinarity that works? What does this imply in terms of specific skills and postures?

The reflection carried out in this chapter to answer these questions has enabled us to review the interdisciplinary experiment conducted at the Université Grenoble Alpes concerning the risks of flash floods since 2003. We propose here to take up the five principles proposed by [BRO 15] to create a successful interdisciplinarity, and to verify how our experience positions itself on each of them.

– Defining a common mission: in our experience, the mission we have set ourselves is to understand what contributes to the impacts of a flood, considered as the interaction of hydrometeorological and social dynamics. Beyond the academic exercise, a shared concern within the team is to provide risk management stakeholders with elements of process understanding that will allow them to better manage crises and reduce the impacts of future events.

– Training interdisciplinary researchers: [BRO 15] speaks of "T-shaped researchers" to illustrate the *oscillation* of junior researchers in interdisciplinarity, between the need to affirm the *primacy of one's discipline* and the hesitant withdrawal from other disciplines. Training should gradually reduce this oscillation to bring researchers into an area of *constructive dialogue* necessary for interdisciplinary construction. The experiments we carried out during the summer schools and the Franco-Hellenic master's course, but also the support of doctoral students involved in our work, new researchers in the framework of the various projects or the AMETIhST seminar, have apparently made it possible to create the conditions for this constructive dialogue and to achieve the production of new knowledge.

– Fostering constructive dialogue: here again, the AMETIhST seminar and the *Water & Society* summer school are privileged places to foster this dialogue. However, the daily work on projects, such as MobiClimEx on social mobility in the face of flash floods, has also provided us with the opportunity to reflect in depth and find useful methodological tools to progress toward the common objective.

– Institutional material: while the interdisciplinary construction was initially carried out solely by the will of some researchers convinced of its usefulness, some institutional material has also made it possible to continue this effort. Among these materials, we can note, for example, the recruitment by the CNRS, in an interdisciplinary section, of a researcher who has contributed to this adventure from the beginning. The university has also enabled the integration of social science researchers into a hydrology laboratory to strengthen existing links, thus accelerating the integration of approaches. Indeed, as we have mentioned, co-presence is an important condition for interdisciplinary construction and making temporary connections possible is a plus.

– Linking research, practice and policy decision making: as mentioned earlier, several of the interdisciplinary research projects we have carried out have involved more or less directly different categories of operational actors: insurers, road managers, crisis managers, urban planning services, and so on.

This assessment, based on the work of [BRO 15], although it lacks precision on the variables and units of measurement to be mobilized, leads us to conclude that our interdisciplinary experiment can be considered successful. In any case, it has enabled us to make thematic progress in understanding the socio-hydro-meteorological processes involved in flash flood events. It has also made it possible to develop methodological tools to facilitate an interdisciplinary approach to environmental issues. These tools are now being used in the context of other issues such as climate change and its impacts on socio-ecosystems, involving other disciplines (the Trajectories project underway at the Université Grenoble Alpes[11]). Based on these new applications, discussions are continuing within the AMETIhST seminar, which this year focuses on the complementarity between the spatio-temporal approach inspired by Holling [HOL 01] and other interdisciplinary tools deployed on biodiversity themes. The interdisciplinary journey is therefore not over and will continue to enrich our understanding of the complex problems of our societies.

1.6. References

[AND 05] ANDERSON J.A., "Does interdisciplinary research require interdisciplinary education?", *The Forestry Chronicle*, vol. 81, no. 6, pp. 785–786, 2005.

[ANT 01] ANTOINE J.-M., DESAILLY B., GAZELLE F., "Les crues meurtrières, du Roussillon aux Cévennes", *Annales de Géographie*, vol. 622, pp. 597–623, 2001.

11 https://trajectories.univ-grenoble-alpes.fr/.

[ASP 08] ASPE C., JACQUE M., L'interdisciplinarité sciences sociales-sciences de la nature et de la vie: un paradigme inachevé, *18ème Congrès de l'Association Internationale des Sociologues de Langue Française*, Istanbul, Turkey, July 7–11, 2008.

[BAR 17] BARTHEL R., SEIDL R., "Interdisciplinary collaboration between natural and social sciences – status and trends exemplified in groundwater research", *PLoS ONE*, vol. 12, no. 1, e0170754, 2017.

[BOI 13] BOISSIER L., La mortalité liée aux crues torrentielles dans le Sud de la France: une approche de la vulnérabilité humaine face à l'inondation, Doctoral dissertation, Université Paul Valéry-Montpellier III, Montpellier, France, 2013.

[BOU 06] BOURQUE L., SIEGEL J., KANO M. *et al.*, "Morbidity and Mortality Associated with Disasters", in QUARANTELLI E., DYNES R. (eds), *Handbook of Disasters*, Springer, Berlin, pp. 97–167, 2006.

[BRI 05] BRINGE A., LAURENT R., *Reconstituer des histoires individuelles à partir de données de suivi démographique*, CEPED, Paris, 2005.

[BRO 15] BROWN R.R., REBEKAH R., DELETIC A. *et al.*, "Interdisciplinarity: how to catalyse collaboration", *Nature News*, vol. 525, no. 7569, p. 315, 2015.

[CAS 06] CASH D., ADGER W.N., BERKES F. *et al.*, "Scale and cross-scale dynamics: governance and information in a multilevel world", *Ecology and Society*, vol. 11, no. 2, p. 8, 2006.

[EVA 06] EVANS R., MARVIN S., "Researching the sustainable city: three modes of interdisciplinarity", *Environment and Planning A*, vol. 38, no. 6, pp. 1009–1028, 2006.

[GIB 94] GIBBONS M. (ed.), *The New Production of Knowledge: The Dynamics of Science and Research in Contemporary Societies*, Sage, London, 1994.

[GIB 19] GIBSON C., STUTCHBURY T., IKUTEGBE V. *et al.*, "Challenge-led interdisciplinary research in practice: program design, early career research, and a dialogic approach to building unlikely collaborations", *Research Evaluation*, vol. 28, no. 1, pp. 51–62, 2019, available at: https://doi.org/10.1093/reseval/rvy039.

[GIE 14] GIEC, Changements climatiques: Incidences, adaptation et vulnérabilité Résumés, foire aux questions et encarts thématiques. Contribution du Groupe de travail II au cinquième Evaluation report by the Intergovernmental Panel on Climate Change [Edited by FIELD C.B., BARROS V.R., DOKKEN D.J. *et al.*], Organisation météorologique mondiale, Geneva, Switzerland, 2014.

[GLE 16] GLEED A., MARCHANT D., Interdisciplinarity, Survey report for the Global Research Council, DJS Research, available at: https://www.jsps.go.jp/english/egrc/data/5th/Survey_Report_on_Interdisciplinarity_for_GRC_DJS_Research.pdf, 2016.

[GRO 07] GROSSETTI M., "L'espace à trois dimensions des phénomènes sociaux", available at: halshs.archives-ouvertes.fr/halshs-00477433, 2007.

[GRU 77] GRUNTFEST E., What people did during the big thompson flood, Working Paper 32, Natural Hazard Center, Boulder, 1977.

[GRU 00a] GRUNTFEST E., "Flash Floods in the United States", in PIELKE J.R., PIELKE S.R. (eds), *Storms*, Routledge, London, pp. 192–206, 2000.

[GRU 00b] GRUNTFEST E., RIPPS A., "Flash floods: warning and mitigation efforts and prospects", [PARKER D. (ed.)], *Floods*, vol. 1, pp. 377–390, 2000.

[HAD 08] HADORN G.H., BIBER-KLEMM S., GROSSENBACHER-MANSUY W. *et al.*, "The emergence of transdisciplinarity as a form of research", in HADORN G.H., HOFFMANN-RIEM H., BIBER-KLEMM S. *et al.* (eds), *Handbook of Transdisciplinary Research*, Springer Netherlands, Dordrecht, pp. 19–39, 2008.

[HOL 01] HOLLING C.S., "Understanding the complexity of economic, ecological, and social systems", *Ecosystems*, vol. 4, no. 5, pp. 390–405, 2001.

[KLE 10] KLEIN J.T., "A taxonomy of interdisciplinarity", in FRODEMAN R., KLEIN J.T., MITCHAM C. (eds), *The Oxford Handbook of Interdisciplinarity*, Oxford University Press, Oxford, pp 15–30, 2010.

[LAK 07] LAKHANI K.R., PANETTA J.A., "The principles of distributed innovation", *Innovations: Technology, Governance, Globalization*, vol. 2, no. 3, pp. 97–112, 2007.

[LED 15] LEDFORD H., "Team science", *Nature*, vol. 525, no. 7569, p. 308, 2015.

[LOO 12] LOOPER J.P., VIEUX B.E., "An assessment of distributed flash flood forecasting accuracy using radar and rain gauge input for a physics-based distributed hydrologic model", *Journal of Hydrology*, vol. 412–413, pp. 114–132, 2012.

[LUT 16] LUTOFF C., CREUTIN J.D., RUIN I. *et al.*, "Anticipating flash-floods: multi-scale aspects of the social response", *Journal of Hydrology*, vol. 541, pp. 626–635, 2016.

[MAR 10] MARCHI L., BORGA M., PRECISO E. *et al.*, "Characterisation of selected extreme flash floods in Europe and implications for flood risk management", *Journal of Hydrology*, vol. 394, nos 1–2, pp. 118–133, 2010.

[MEN 12] MENONI S., MOLINAR D., PARKER D. *et al.*, "Assessing multifaceted vulnerability and resilience in order to design risk-mitigation strategies", *Natural Hazards*, vol. 64, no. 3, pp. 2057–2082, 2012.

[NAT 15] NATURE, "Why interdisciplinary research matters", *Nature*, vol. 525, no. 7569, p. 305, 2015, doi: 10.1038/525305a.

[NOW 03] NOWOTN H., "Democratising expertise and socially robust knowledge", *Science and Public Policy*, vol. 30, no. 3, pp. 151–156, 2003.

[NOW 06] NOWOTNY H., SCOTT P., GIBBONS M., "Re-thinking science: mode 2 in societal context", in CARAYANNIS E.G., CAMPBELL D.F.J. (eds), *Knowledge Creation, Diffusion, and Use in Innovation Networks and Knowledge Clusters: A Comparative Systems Approach across the United States, Europe and Asia*, Praeger, Westport, CT, pp. 39–51, 2006.

[PET 08] PETT J., OWEN S., BULKELE H., "Crossing boundaries: interdisciplinarity in the context of urban environments", *Geoforum*, vol. 39, no. 2, pp. 593–601, 2008.

[PIC 14] PICARD, E., "Disciplines académiques et cadre institutionnel: réflexions autour du cas français", Balz Engler. Disziplin/discipline. *28 Kolloquium der Schweizerischen Akademie der Geistes- undSozialwissenschaften*, Academic Press, pp. 47–56, 2014.

[RUI 07] RUIN I., Conduite à contre-courant. Les pratiques de mobilité dans le Gard: facteur de vulnérabilité aux crues rapides, PhD thesis, Université Joseph-Fourier-Grenoble I, Grenoble, 2007.

[RUI 09] RUIN I., CREUTIN J.D., ANQUETIN S. *et al.*, "Human vulnerability to flash floods: addressing physical exposure and behavioural questions", *Flood Risk Management: Research and Practice Proceedings of the European Conference on Flood Risk Management Research into Practice (FLOOD risk 2008)*, Oxford, UK, September 30–October 2, 2008, pp. 1005–1012, 2009.

[RYL 15] RYLANCE, R., "Grant giving: global funders to focus on interdisciplinarity", *Nature News*, vol. 525, no. 7569, p. 313, 2015.

[RUI 14] RUIN I., LUTOFF C., BOUDEVILLAIN B. *et al.*, "Social and hydrological responses to extreme precipitations: an interdisciplinary strategy for post-flood investigation", *Weather, Climate and Society, Weather, Climate & Society*, vol. 6, no. 1, pp. 135–153, 2014.

[SCO 12] SCOLOBIG A., DE MARCHI B., BORGA M., "The missing link between flood risk awareness and preparedness: findings from case studies in an Alpine Region", *Natural Hazards*, vol. 63, no. 2, pp. 499–520, 2012, doi: 10.1007/s11069-012-0161-1.

[SHI 78] SHINN T., "Des Corps de l'Etat au secteur industriel: genèse de la profession d'ingénieur, 1750–1920", *Revue française de sociologie*, vol. 19, no. 1, pp. 39–71, doi: 10.2307/3320953, 1978.

[SNO 64] SNOW C.P., *The Two Cultures: And a Second Look*, Cambridge University Press, Cambridge, 1964.

[SPE 10] SPERBER D., "Pourquoi repenser l'interdisciplinarité", *Repenser l'interdiscplinarité*, virtual seminar, available at: www.interdisciplines.org, 2010.

[STA 94] STAES C., ORENG J.C., MALILA J. *et al.*, "Deaths due to flash-floods", *Puerto-Rico, January 1992: Implications for Prevention. International Journal of Epidemiology*, vol. 23, no. 5, pp. 968–975, 1994.

[VIN 11] VINET F., BOISSIER L., DEFOSSEZ, S., "La mortalité comme expression de la vulnérabilité humaine face aux catastrophes naturelles: deux inondations récentes en France (Xynthia, Var, 2010)", *[VertigO] La revue électronique en sciences de l'environnement*, vol. 11, no. 2, 2011.

[YEG 15] YEGROS-YEGROS A., RAFOLS I., D'ESTE P., "Does interdisciplinary research lead to higher citation impact? The different effect of proximal and distal interdisciplinarity", *PLoS ONE*, vol. 10, no. 8, e0135095, 2015.

2

The Pace of Urbanization in Flood-Prone Areas

2.1. Introduction

Urbanization in floodplains[1] is a reality [SCA 14] and a diachronic study [BOU 18] has shown to what extent floodplains have been developed. This chapter proposes a complementary approach aimed at continuously reporting (year by year) the construction of these zones, to capture their paces and trends over the long term (1950–2012). Urbanization is carried out according to a regulatory framework based on national and local standards in a particular territorial context (land pressure, economic growth, agricultural decline, etc.). The artificialization of soils reflects the balance of power between actors who are sometimes far removed from the environmental challenges they have to face. We are also investigating whether the urbanization of floodplains is sensitive to the occurrence of extreme hydrometeorological phenomena. After catastrophic events (warnings, rescue organizations, etc.) [RUI 08], does the public authority change its practices? If so, according to what time frame, and according to what modalities?

The hypothesis supported here is that extreme hydrometeorological phenomena unbalance the production of buildings in areas likely to be flooded. We assume that the downward variations in the number of

Chapter written by Sylvie DUVILLARD, Isabelle ANDRÉ-POYAUD and Charles-Antoine CHOQUET.

1 See Box 2.2 on PPRI's, *plans de prévention au risque d'inondation* (French flood risk prevention plans).

constructions in floodplains reflect the consideration of risk by decision-makers.

Our analysis is carried out at different spatial and temporal scales, building on previous work on the relationship between construction and flooding (first part). While they tend to show in part the impact of floods on urbanization, it will be a question of assessing their temporalities (second part) and local particularities (third part). To do this, we will propose an empirical analysis based on data from tax authorities' land files[2], a study conducted in the Gard department in southeastern France.

2.2. The pace of urbanization and flooding: a difficult correlation to be established

The work of Scarwell [SCA 14] and Laganier [LAG 06] analyzed the consideration, in urban dynamics, of flood risk; both through the evaluation of public policies implemented at different scales [LAG 06], and through the tools available to decision-makers to make the protection of populations operational in the face of sudden water rises [BON 16], [SCA 14]. The priority is to observe the way in which local authorities deal with the issue of risk or not, even though they are obliged to do so (national legislation) and the means (decentralization obliges them). E. Moulin [MOU 14] then questions the strength of the regulation produced by the legislator to constrain the urban form defined by local actors. After having analyzed the process of applying the standard, a derogatory regime then emerges insofar as the "territorialized standard embodies an acceptable risk" [MOU 14]. In the face of risks, the adaptation of national policies and standards to local specificities reflects their necessary adequacy with "social, cultural, institutional, economic, and geographical contexts" [LAG 06]. Are there other factors involved? Which are they in this case?

2.2.1. *Flooding, urbanization and real estate markets: complex interactions*

N. Meschinet de Richemond [MES 06], taking up previous work on "flood-proof" building permits [SCA 14] notes that there has been no

2 Files from the Dgfip 2013 manufacture year.

significant decrease in the number of building permits[3] granted following the entry into force of new laws[4]. The case studies here highlight the inconsistency of local policies and the disengagement of the State by disconnecting the logic of urbanization from that of protection [VAL 09]. The political stakes are particularly complex in a system where decentralization gives local elected officials "responsibility for the implementation of urban planning law supposed to allow the application of preventive measures" even though the central actor is the State [LAR 15]. Authors [LAR 15] debate the depoliticization of the risk issue at a time when the question of responsibility is being judged, particularly following the Xynthia storm in France in 2010, which killed 27 people. In their view, the decentralization and inconsistency of local elected officials mask both the State's disengagement from the prevention system in favor of its role as protector (CatNat decree) and "the limits of existing legal instruments" in the face of the territorial integration of risk relations.

These results suggest that floods are poorly understood at local level, leaving urbanization to follow a different logic (that of the market, for example) than that of taking risk into account. For Douvinet *et al.* [DOU 11], "the absence of serious events for several years contributes to the forgetting the existence of risk, which tends to no longer be integrated into the decision-making process inherent in development".

Approaching pace, in this case, provides a complementary interpretation; economists who use econometric models to study the effects of floods on the housing market have obtained nuanced, sometimes contradictory results. Observing the flood/real estate price relationship, some, like Kouski [KOU 10], deduce the absence of a direct impact on prices, unlike insurance premiums, which increase accordingly. Yet they point out the difficulty of disentangling post-flood changes year by year. Others, like Mauroux [MAU 15], who conducted a literature review on the effect of natural disasters on property prices in floodplains, pointed out that prices actually fell immediately after disasters [HUB 03], [MAU 15]. In the same study, Mauroux [MAU 15] mobilized in particular the work of North American

3 In our opinion, the work on building permits has certain biases: lack of reliable information; gap between the permits granted and the completion of the construction. On this subject of the data relevant for the analysis, see Chapter 8 of this volume.

4 The adoption of the French Law on Compensation for Natural Disasters in 1982, implemented by the Decree in 1984, and the Barnier Law of 1995 on the Protection of Persons and Property in the Event of Flooding and Other Natural Hazards.

economists [MAU 15] where the authors proposed theoretical models on the changes in house prices over time, and their porosity toward natural disasters in the long term. From this work flows our analysis of the trends in urbanization and its paces in flood-prone areas.

The use of these studies lies in the correlation established between decision-making according to the level of theoretical information (rational agents in a situation of perfect information versus a situation of "myopia and amnesia")[5] and the rates at which events occur. They assume that the more risks multiply and get closer in time, the more property prices fall in the long term in the affected areas. The forgetfulness is short-lived, with time steps between each event thwarting a return to high price levels in affected areas.

Without implementing econometric models, we wanted to report on the interactions between the most significant flood dates and the change in urbanization in floodplains between 1950 and 2012 at several spatial scales (on the scale of the Gard and the commune). Kouski [KOU 10] studied price trends after a flood disaster in Saint–Louis (Mississippi) and explained changes in property prices as signs of a risk culture and its "capitalization[6]" by stakeholders (buyers of properties in flooded or flood-prone areas). We consider like him that reducing openness to urbanization reveals that risk is taken into account by political decision-makers. Its results also highlight the variability of prices according to the territorial context[7], so we will draw up a typology of the study municipalities according to their exposure to risk, thus taking into account the demographic (size of the municipality and the changes in its population) and spatial (communal area exposed to risk) variables.

2.2.2. *The data mobilized: the number of constructions recorded in tax files*

In this section, we propose to describe the data used to analyze the changes in urbanization in flood-prone areas by comparing it with that deployed in non-flood-prone areas. First, we will present the statistical

5 Disappearance of the risk memory.

6 Actors in the real estate market integrate flood risk following a succession of events occurring closer and closer in time.

7 He then observed whether learning about flooding varied across the Municipalities of Saint Louis County, Mississippi, still considering changes in property prices as significant for the level of information provided to real estate market participants.

sources used to work on urbanization, and then the geographical data that are used to define the envelopes of floodplains, bearing in mind that all the analyses presented in this chapter are based on a succession of methodological choices that have repercussions on the number of buildings observed in each area.

In France, it is possible to mobilize different sources of geographical data to analyze the phenomenon of urbanization as a whole or according to particular prisms, such as the comparison of urbanization in flood-prone areas versus non-flood-prone areas. A first approach consists of comparing a succession of maps of a territory at different dates, as proposed by Boulier *et al.* [BOU 15], who used a set of 1/25,000 IGN digital maps dated from 1979 to 2011, as well as the data from the 2012 BD Topo.

Based on the cartographic work on these sources, these authors evaluated the end construction date[8] of each identified building, as well as the nature of its distinguished use in agricultural, commercial, industrial, noteworthy, or undistinguished buildings (of the habitat or mixed building type). The number of buildings was then calculated by using the construction period, the risk level, and the urbanization area (urbanizable or not). The data used in this approach make it possible to accurately estimate the stock of buildings by type over each long-term period, including mixed buildings that include dwellings. However, they do not provide information on the type of housing, whether individual or collective, or on the number of dwellings recorded in each building, or on the nature of the dwellings (HLM or not) or their type of occupation (primary or secondary residence). The land data from the General Directorate of Public Finance (DGFIP) provide access to this information[9]. By providing homogeneous data on plots, premises, and their owners for the whole of France (including the French overseas departments), this rich but complex source of information opens up interesting working prospects for territorial studies, particularly on land use and planning issues. On the other hand, the finesse of information, which can be gained from the description of buildings and premises, cannot be preserved over time. Indeed, Cerema[10], which disseminates these data to

8 Five construction periods have been defined: before 1979, from 1979 to 1991, from 1991 to 2002, from 2002 to 2007, and from 2007 to 2012.

9 However, we will not cover all of this information in this study.

10 Cerema (*Centre d'études et d'expertise sur les risques, l'environnement, la mobilité et l'aménagement*) is a French public institution under the dual supervision of the Ministry of Ecological and Solidarity Transition and the Ministry of Territorial Cohesion. Through its

local authorities and State structures, such as DDTM[11], has only had access to the dates of the land files since 2009.

It is from the data in the land files (see Box 2.1 "land files") that we present in this chapter a complementary work to that of Boulier *et al.* [BOU 15] by relying not on the building as a whole but on a finer entity, that of the local area. A room can be of four different types: a house, an apartment, an industrial or commercial space, and an outbuilding.

To this end, we have chosen to use the 2013 manufacture year of the land files updated on January 1, 2013, by retaining residential or business premises, which makes it possible to account for human exposure to flood risks both in residential areas and in some of the places where daily activities are carried out[12]. To describe the premises, we have chosen their year of completion of construction or fitting-out in an existing building, and their geographical location defined from the parcel in which they are referenced. The outline of buildings and premises is not defined as such, but it is possible to know the spatial extent of the parcels or land units to which they are attached. A land unit is defined as a group of adjoining parcels of land located in the same municipality that belongs to the same owner or to the same joint ownership. In the case of land units, all buildings are attached to a reference parcel, most often the smallest, even if they are physically located on different parcels. This lack of spatial accuracy has methodological consequences on the allocation of premises in flood-prone or non-flood-prone areas and from a risk level to the risk of flooding if necessary.

In addition, we encountered topology problems on the plots, particularly during geographical processing (spatial joint) between the plot layers and PPRI zoning, because the computerized cadastral plan of the Gard was not yet fully digitized in vector form. We preferred to relate the plot characteristics to the geometric data of the IGN BD Parcel. Unfortunately, we had one for 2011 and not 2013. While this methodological choice made it

expertise, it aims to support territorial actors in the implementation of their public policies, particularly on issues concerning territorial planning and cohesion, the environment, risk prevention, mobility and transport, and well-being (https://www.cerema.fr/fr).

11 DDTM: *Direction départementale des Territoires et de la Mer.* The French Departmental Directorate of Territories and the Sea. A decentralized State service in the French departments.

12 Even if, unfortunately, these places of activity are underestimated insofar as the land files only take into account business premises paying a business tax. It should be noted that public buildings not subject to property tax are not included in this source.

possible to improve the geometric accuracy of the plots, we had to deal with problems related to the time lag between the two tables. Parcel identifiers from the land files did not find a match in the BD Parcellaire. In total, we counted 7,702 premises without plot allocation out of the 597,871 premises recorded in the Gard (i.e. 1.2%).

The land files come from the MAJIC application (updated cadastral information) of the DGFIP (*direction générale des finances publiques*). They describe the built and unbuilt properties subject to tax, as well as the owners of these properties. Cerema Nord–Picardie ensures the distribution of these files and offers pre-processing to facilitate their exploitation.

Structuring

The six DGFIP raw files (owner files, built property file, undeveloped property file, batch property file, batch and local correspondence file, and the Fantoir file) are reworked into a postgre/postgis database. This database, consisting of 16 tables, contains the initial land files restructured into several tables to which are added aggregated tables that contain the same initial data aggregated according to different units. These include, for example, tables on buildings and land units at different levels of scale.

Since 2009, Cerema has been proposing a new manufacture year for each year (year N) based on the data available in the MAJIC application on January 1 of year N. The use of these files deserves clarification on the latency issues of this file. In all manufacture years of year N, new buildings dating from year N – 1 are underestimated due to the time required to update them. Premises not registered in the manufacture year of year N will be registered in the following manufacture year.

Definition of developed and undeveloped properties (parcels)

Developed properties are considered to be housing, its annexes, and business premises subject to business tax. Undeveloped properties are identified by parcel. As the data from this source are intended for tax calculation purposes, the areas corresponding to public domain rights-of-way are not recorded in these files. For example, for unregistered areas of roads, state forests, military land, parks or public gardens, or water surfaces. The same applies to all public buildings such as schools and hospitals, which are tax exempt. On the other hand, public buildings' housing staff are listed in this data source.

Parcels

Cadastral parcels are portions of land in one piece that belong to the same owner or to the same undivided co-ownership. Located in the same commune and in the same cadastral section, they are identified by a unique number throughout France (idpar), which corresponds to the juxtaposition of the department code, the cadastral section number, and a parcel number. Adjacent parcels of the same municipality may constitute a land unit if they belong to the same owner or to the same joint ownership.

Geolocation of parcels of land

In land files, parcels are defined as geographical objects and have a polygon-shaped spatial hold. This geolocation is carried out by Cerema. Until 2013, it used only the computerized cadastral plan, digitized in vector form, but encountered difficulties with incomplete computerized plans where spatial rights-of-way took a simplified square-type form. These simplified shapes created problems in calculating surfaces that could be biased and topology problems also, due to the simplified outlines overlapping the outlines of correctly digitized parcels.

The premises

Buildings can be composed of one or more rooms. A premises is defined as a set of different buildings that form a homogeneous real estate complex belonging to the same owner. A house with an insulated garage (not attached to the dwelling) on a plot forms a lot. On the other hand, if this garage is separated from the house by a road, then two premises are counted (CEREMA NORD-PICARDIE, 2017, a guide for handling land files). In an apartment building, as many rooms are registered as there are separate apartments.

The premises are characterized by a large number of descriptors, including their physical form, identified in four types: houses, apartments, outbuildings (buildings isolated from any main residential area), or business premises. Generally speaking, an activity room only includes activity. In the context of a professional activity of a liberal type (doctor's office for example) carried out in a main dwelling, the DGFIP distinguishes two premises, one of activity and the other dwelling. As part of our work, we have mobilized this variable related to the physical nature to count the number of premises. We also used the year of completion of construction or completion of work (Janath) to trace the history of construction by type of premises. Necessary for calculating taxes, this information has been collected since 1970 for residential housing and since 2002 for business premises. This variable, which is updated regularly, first provides the year in which a room was completed. On the other hand, if the premises undergoes significant restructuring work, the year of completion of the work is substituted for the latter.

Location of the premises

Each premises is attached to a plot. It is identified by a unique number throughout France, consisting of the department and a local number. In this source, unlike parcels, the spatial hold of a room (or a building as a whole) is not precisely defined. The premises are only located at the plot by a point object with an x and y coordinate.

Box 2.1. *DGFIP's land files*

2.2.3. *Mobilizing PPRI zoning to assign a floodplain and a risk level to a plot and a premises*

To identify the spatial extent of floodplains in the Gard, several data sources were combined. These are the zonings defined as part of the flood risk prevention plans (PPRI) (see Box 2.2 "PPRI") and made available on the DDTM du Gard website.

Flood risk prevention plans (PPRIs) are instituted and made mandatory under the "Barnier" law of February 2, 1995, No. 95–101 on strengthening environmental protection. They replace the *plans de surface submersible*, PSS, submersible surface plans created by decree in October 1935, and the *plans d'exposition aux risques naturels prévisibles*, PER, exposure plans for foreseeable natural hazards established with the law on compensation for natural disasters of July 13, 1982, and its implementing decree of May 1984. Established under the authority of the prefect after consulting the municipal councils of the municipalities concerned and conducting a public inquiry, these regulatory documents aim to delimit areas exposed to the risk of flooding of persons and property. They provide for a ban on new construction in the most exposed areas and a strong constraint on the conditions for the construction and use of new buildings in less exposed areas. This tool, which regulates land use in high-risk areas, is integrated into municipal urban planning documents and consulted when building permits are issued. The zoning of PPRI is based on reference data from archives and historical or centennial flood records.

Box 2.2. *Flood risk prevention plans*

The zoning of the first PPRI, known as "old-generation PPRI," i.e. approved before 2002, was often considered at the scale of watersheds and therefore included several municipalities. They proposed two types of zones: low-risk zones where new construction was allowed under specific development conditions (e.g. raised ground floor, crawl space) and high-risk zones where new construction was prohibited with few exceptions. With the 2003 law on the prevention of technological and natural hazards and the repair of damage, these PPRIs have evolved into so-called "new-generation" PPRIs. The evolution lies in the category of zoning with the proposal to distinguish different risk levels that reflect different levels of exposure. Three risk levels are thus identified: residual risk areas, moderate risk areas, and high-risk areas. In addition, there may be a fourth level, very high risk, as is the case for Nîmes. At the end of 2016, there were 51 PPRIs in the Gard region, 42 of them new-generation and nine old-generation[13].

It is from these different sources that we have reconstructed the PPRI zoning of each municipality. This work required GIS geoprocessing manipulations to preserve for each municipality the zoning of the last PPRI in force. In fact, we have excluded all municipalities with an approved new-generation PPRI from the zoning of first-generation PPRIs still in force. At the end of 2016, in the Gard, 223 municipalities had a PPRI, the dates of approval of these PPRI varying between 1995 and 2016. Figure 2.1 shows the PPRI zoning on which we based our study. It appears that 28% of the area of the Gard is identified as a floodplain.

The fact of combining PPRI zoning from two generations raises the problem of the different classifications used to qualify the risks. In an attempt to harmonize these classifications, we have established a correspondence between the risk levels of first-generation PPRIs and the risk levels of the latest generation PPRIs using the following rules. The highest risk zoning is associated with the high risk level. High-risk or low-risk zonings correspond to the moderate risk level. Old-generation PPRIs, such as those of Rohny or the Rhône Cèze Tave confluence, had identified a "low risk" level. We have associated the latter with the level of residual risk. In the operations carried out, we refer to the use of this correspondence table only to describe the stock of buildings by risk level, but in most of the processes presented here, the stock of buildings is analyzed according to its location in flood-prone or non-flood-prone areas. The "flood-prone areas" include the spatial control of all risk levels defined in the new generation of PPRIs (i.e. residual, moderate, strong, and very strong for Nîmes).

13 Old generation PPRI include: the lower Arre, Gardon Anduze, downstream Gardon, high Vidourle, Rohny, Le Rohnu, middle Vidourle, middle Vistre, Rhône Cèze Tave confluence, and the Rhône Gardon Briançon confluence.

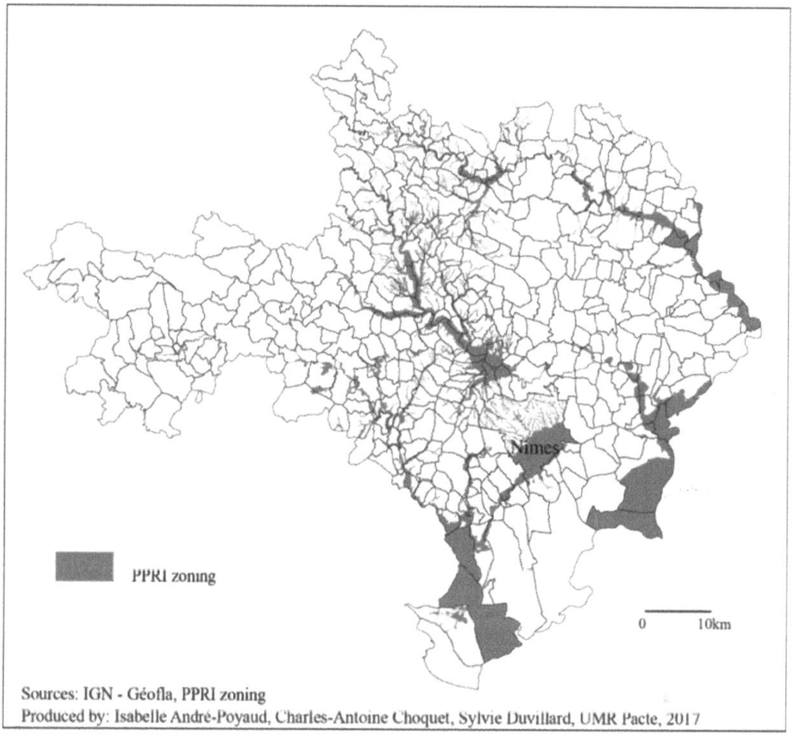

Figure 2.1. *Map of the PPRI envelope in the Gard*

In order to define the floodability or not, then the risk level, of a parcel and the premises located there, we proceeded in stages. First, we divided each plot into sub-polygons based on the spatial rights-of-way of the risk levels, and then calculated the area exposed to each risk level. We then calculated the percentage of the parcel area located in a flood-prone and non-flood-prone area. When more than half of the plot's surface area had a zero risk, the entire plot was then defined as a non-flood-prone area. In the case where more than half of the plot was located in low-, medium-, high-, or very high-risk areas, the plot was affected by the risk level corresponding to the most significant area. This criterion of choice fixed according to the majority surface area seemed more relevant to us than an excessive precautionary principle, which would have consisted of affecting the highest level of risk without taking into account the surface affected by the risk level. The plot's risk level was then passed on to all the premises located on it. Using this method, we evaluated 297,839 premises located on plots defined in non-flood-prone areas (64%) and 166,472 premises on plots defined in flood-prone areas, of which 28% were residual, 36% low risk, 30% high risk, and 6% very high risk.

2.3. Urbanization rates in flood-prone areas in the Gard department

The pace analysis was carried out according to several steps. After observing the rate of construction per year over time, we then tried to assess them according to the types of construction (houses, apartments, and business premises); at the same time, we introduced the dates of major floods [TRO 04][14], i.e. those affecting the whole of the Gard (1958, 1988, and 2002), and those marked by the implementation of national regulatory constraints (1984 and 1995[15]). It then interpreted, in part, the capacity of the actors to take into account the flood risk after a major hydrometeorological event [KOU 10], after a succession of disasters [MAU 15], [PRY 11] or under the constraint of a standard.

2.3.1. *Strong inter-annual fluctuations and a significant reduction in urbanization in floodplains from 1995 onward*

Figure 2.2 shows, based on the stock of premises available at the end of 2012 in the land database, i.e. built and still present on the territory, the dynamics of urbanization in the Gard between 1950 and 2012 and the location of these premises in flood-prone areas or not. At the end of 2012, the Gard had 605,382 apartments, houses, and business premises, more than three-quarters of which (464,311) were built between 1950 and 2012. These included 239,387 houses (51.6%), 187,239 apartments (40.3%), and 37,685 commercial or industrial premises (8.1%)[16].

The first part of Figure 2.2 shows the number of premises, regardless of their nature (apartments, houses, and business premises), by year of completion (or completion in the case of regeneration) between 1950 and 2012. These constructions followed the main phases of the French economic

14 The authors analyze the impact of California's AB 1196 on house prices (hedonic price method) without being able to assert the significance of the drop in value.

15 The adoption of the Law on Compensation for Natural Disasters in 1982, implemented by decree in 1984, and the Barnier Law of 1995, on the protection of persons and property in the event of floods and other natural hazards.

16 Accepting the bias that before 1970 the information in the DGFIP database is incomplete.

cycle of recent decades and went hand in hand with the change in the population of the Gard, which increased on average by 1.04% per year and almost doubled in 60 years, from 396,742[17] inhabitants in 1954 to 733,201 inhabitants in 2013[18]. The second part of Figure 2.2 compares for each year the proportion of premises built in flood-prone and non-flood-prone areas according to the PPRI zoning in effect at the time of the study (2015). While urbanization in floodplains was more sustained during the period preceding the first oil crisis (the proportion of premises in floodplains reached nearly 70% in 1961), it tended to decline slowly from 1990 onward without being permanently abandoned. Since the early 2000s, each year, nearly 30% of residential and business premises have been built in flood-prone areas of the Gard.

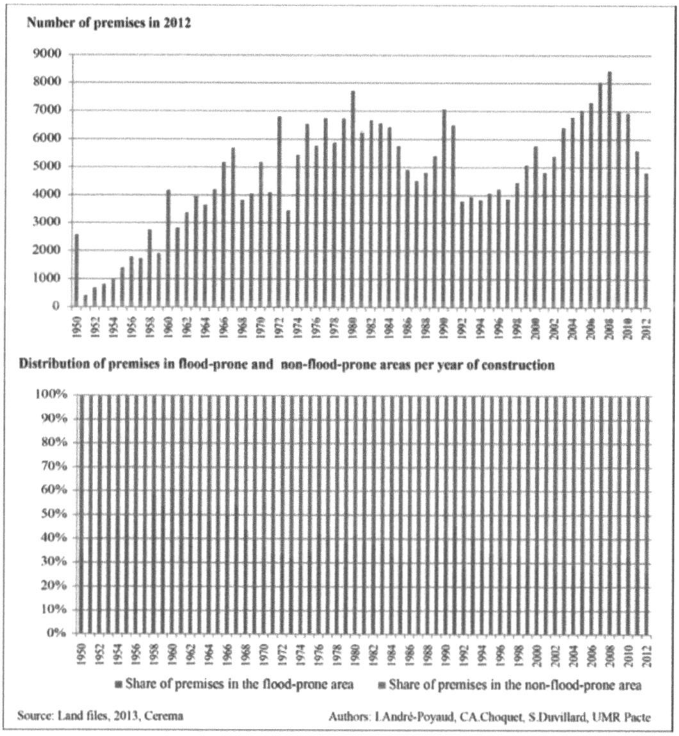

Figure 2.2. *Distribution of residential and business premises available in 2012 in the Gard according to their year of construction and location. For a color version of this figure, see www.iste.co.uk/lutoff/hydrometeo2.zip*

17 Source: https://fr.wikipedia.org/wiki/D%C3%A9mographie_du_Gard#cite_note-splaf-2.

18 Source: INSEE, 2013, population census.

This general downward trend observed on a Gard scale for all premises masks differences, particularly according to the level of exposure of the municipality to the risk of flooding and the size of its population. Variations also exist according to the type of premises. We will come back to these differences in the following sections. From a global perspective, to remain on the scale of the Gard, we examined whether the major floods experienced in the Gard in 1958, 1988, and 2002 had a direct impact on this downward trend. Similarly, we wish to report on the impact of the 1982 Natural Disaster Compensation Acts and the 1995 Barnier Act.

With Figure 2.3, we focus on residential premises containing houses and apartments in regards to the Gard. For each area (flood-prone or non-flood-prone), the aim is to report on each year's contribution to the stock of available premises.

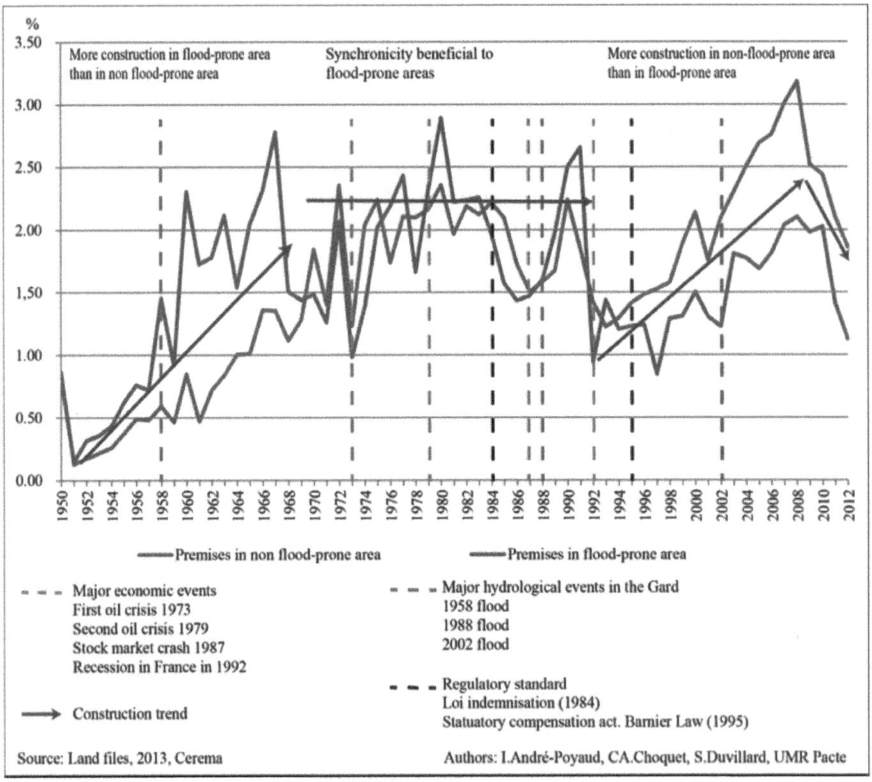

Figure 2.3. *Contribution of completion years to the stock of 2012 residential premises built in flood-prone and non-flood-prone areas. For a color version of this figure, see www.iste.co.uk/lutoff/hydrometeo2.zip*

Over these 62 years of observation, excluding annual variations, it is possible to report three periods. The first period, between 1950 and 1969, showed a more sustained housing stock construction activity in flood-prone areas than in non-flood prone areas. Housing built during this period represented 26.5% of the 2012 housing stock located in flood-prone areas compared to 14.1% of the stock located in non-flood-prone areas. The second phase, which took place between 1970 and 1991, marked a period of dense construction, regardless of the area. The weight of this construction period in the 2012 housing stock was very close in both zones: 42.7% of homes built in non-flood-prone areas and 41.7% of homes built in flood-prone areas. The third period between 1992 and 2012 was marked by a reversal of the trends observed during the first period with much more construction activity in non-flood-prone areas than in flood-prone areas: 43% of the housing stock in non-flood-prone areas has been completed over the last 20 years compared to 31% in flood-prone areas. In addition to this general trend, the graph also shows large variations in construction from one year to the next. The rate of construction reacts over a short period of time to external factors of any kind, but we are looking at whether these events have a long-term impact. To do this, we have positioned two main types of events on the graph. These are major economic and financial disruptions and floods that have impacted the Gard. For economic disruptions, we used the two oil crises of 1973 and 1979, the 1987 stock market crash, as well as France's entry into recession in 1992 and the financial and economic crisis of 2008. For hydrological phenomena, we reported the years of major floods in the Gard such as 1958, 1988, and 2002, as well as the legislator's response with the adoption of the law on compensation for natural disasters in 1982, implemented by decree in 1984, and the Barnier law of 1995 on the protection of people and property in the event of floods and other natural hazards.

All the economic downturns mentioned earlier have had an impact on the construction sector and resulted in a smaller contribution of these years to the housing stock observed in 2012. It should be noted that the decreases were more pronounced in flood-prone areas than in non-flood prone areas. The largest decline (by adding the decreases observed in flood-prone and non-flood-prone areas) was recorded during the first oil crisis of 1973; the second oil crisis seemed to have been better cushioned, at least for homes located in flood-prone areas. While the 1987 stock market crash did not seem to have any short-term impact on the pace of construction in the Gard, the latter was severely affected in the year following the first Gulf War and France's entry into recession at the beginning of autumn 1992. As during the first oil crisis, the decline in activity in the construction sector was felt more in flood-prone areas than in non-flood-prone areas. The stock market and financial crisis of 2008 also disrupted the number of buildings delivered in 2009, with a steeper decline in construction in non-flood-prone areas than in flood-prone ones. In light of these observations, it is possible to establish a downward trend in the pace of construction when it is impacted by exogenous economic factors.

However, it is more difficult to account for this type of relationship between the rate of construction and large-scale hydrological phenomena. Declines in construction are not directly noticeable the year after the impact, but declines can be recorded 2–3 years later[19]. Except for the 1958 flood, the rate of construction reacted with a decrease in the year following the event and another three years later. For the 1988 and 2002 floods, the two years following these rises in water levels were marked by a more pronounced upward variation for buildings located in flood-prone areas than those in non-flood-prone areas. These increases were probably due to the reconstruction work required for some buildings after a natural disaster event. It should be recalled that in land files, the year of renovation is substituted for the year of completion of construction when a building is affected by major renovations such as the need to comply with regulations. The pace of construction seems to be impacted briefly and within a 2–3 year time lag by these hydrological phenomena, but is perhaps primarily part of economic cycles whose impacts are more pronounced. During the 1992 recession, the fall in construction was more severe in flood-prone areas than in non-flood-prone areas. On the other hand, the years following the 2002 flood were part of an upward cycle of construction that continued until 2008. This was mainly due to favorable financial conditions for households, such as lower housing interest rates or longer borrowing periods, which encouraged the purchase of new properties. In addition, there were also tax relief measures for the Robien scheme[20], which supported new housing construction dedicated to rental in medium-sized French cities. Everything suggests that the trend in the economic cycle at the time of the impact of a flood predominates over the choices to maintain (upward cycle) or slow down over a short period of time (downward cycle) the construction in floodplains.

19 Kouski [KOU 10] points to the same time lag in studying price developments year by year following a disaster.

20 Tax exemption scheme set up in France to encourage the construction of new housing.

On the other hand, the comparison of these two time series with the decree implementing the law on compensation for natural disasters in 1984 and the date of approval of the Barnier law in 1995 makes it possible to identify changes in construction policy imposed by the regulatory framework. For the three years following the decree implementing the 1982 law in 1984, the influence of these years in the 2012 stock of premises located in flood-prone areas was steadily decreasing, while the influence of these same years was increasing in the stock of premises located in non-flood-prone areas. However, it was not until 1994 that a lasting reversal in the construction trends between the two areas was observed at the Gard scale. The contributions of the years 1994–2010 to the 2012 stock of premises were significantly higher in non-flood-prone areas and conversely, the contributions of the same years to the 2012 stock of premises located in flood-prone areas were significantly lower. We see here the impact of the regulatory standard aimed at reducing people's exposure to their new places of residence. Despite all this, urbanization in flood-prone areas has not been completely interrupted since it remains possible in the least exposed areas and a significant proportion of the stock of residential premises built before 1994 remains subject to land flooding.

2.3.2. *Discrepancies in paces according to building types*

Land management is an eminently political matter; choices exist, as well as room for maneuver, within a tight framework: on the one hand, the pursuit of development (economic and demographic), and on the other, the electoral imperative where the "owner-elector" has a major impact on the process of opening up to urbanization, while the normative framework narrows the field of possibilities[21]. So how do we interpret these trade-offs, how do the actors use or create room for maneuver to reconcile these different imperatives? Especially in France, the ALUR law[22] partly contradicts the Barnier Law (1995) in the sense that it abolishes the minimum density surface allowing parcel divisions with buildings, including in flood-prone areas[23]. It is then possible to densify in areas subject to risks except in areas with high risk

21 "We have always built in a floodplain by cronyism but for the last 4/5 years it's finished," extract from an interview with a technician, as part of the ANR Mobiclimex, Gard 2017.

22 Act No. 2014–366 of March 24, 2014 on access to housing and renovated urban planning, known as the ALUR Act or the Duflot II Act, is a French law on housing.

23 Provided that the height of +50 cm above the flood markers is respected.

levels[24], as shown in Figure 2.4. This concerns the municipalities that have a PPRI. It cross-references risk levels with building types, showing at first interpreting the redistribution of uses over time and their sensitivity to major events that have affected the department. If the number of buildings, whatever their use (work/housing) or form (house/apartment), decreases after major events and then increases a few years later, the impact on the houses is less marked: the number of houses slowly increased again until 1995 and 2002 and then decreased regularly until 2012. In contrast, the proportion of apartments and business premises in flood-prone areas highlights trade-offs in terms of exposure: more than half of the total number recorded per year is located in flood-prone areas. The investigation by interviews with decision-making actors[25] accompanies the analysis of cadastral data, shedding a little more light on these observations: "companies look at the risk of fire in the short term," especially even if, since 2012, "the State has been focusing intently on the areas of activity where there is still a radiology practice, telecommunications, supermarkets, etc".

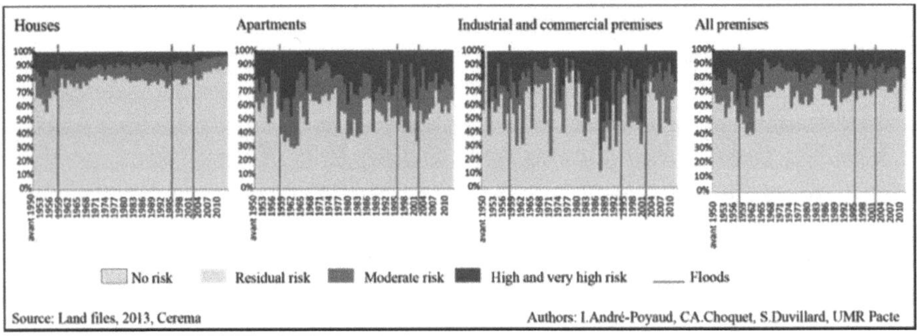

Figure 2.4. *Distribution of premises by risk level between 1950 and 2012 in Gard municipalities with a PPRI. For a color version of this figure, see www.iste.co.uk/lutoff/hydrometeo2.zip*

The juxtaposition of these graphs shows the political choices made over time and the desire to reduce the exposure of detached homes compared to other premises such as apartments or business and industrial premises. Indeed, in these municipalities subject to PPRI, the vast majority of houses have been built in non-flood-prone areas. Despite this, at the end of 2012,

24 Extension limited to 20 m^2, only once.

25 Actors met: elected officials (4), technicians (4), lawyers (3), real estate agents (1), and insurers (1); carried out in the Gard in 2017 as part of the ANR Mobiclimex.

16,397 houses were located in high and very high-risk areas for Nîmes exclusively, which represents 10% of the total number of houses in municipalities with a PPRI. These grounds are characterized as being rather old, with two-thirds of these dwellings built between 1950 and 1970, a period of heavy construction during which politicians were more concerned with access to housing than with the risk of flooding. Nevertheless, over the period studied, there was a gradual decrease in house construction in the areas with the highest risks. This gradual decline in house construction in the most exposed areas was in line with the commitment to a national risk prevention policy that has been in place since the mid-1990s. Between 1950 and 1969, this section accounted for 12% of the individual housing stock of these municipalities, compared with 9% between 1970 and 1995 and 6% after 1995. On the other hand, the proportion of houses built in moderate risk areas remained almost constant over the entire observation period. The same was true for the proportion of houses built with a residual risk level. These percentages were respectively around 10% for the three periods: 1950–1969, 1970–1995, and after 1995.

While the municipalities subject to PPRI tend to reduce the exposure of detached homes, they continue to build residential housing of the apartment type and business or industrial premises in flood-prone areas. In 2012, the number of apartments built in these areas amounted to 97,582, of which 7% had high and very high risk and 29% had moderate risk levels. This distribution remained more or less the same over the three periods 1950–1969, 1970–1995, and 1996–2002. The trajectory of business and industrial premises was similar to that of apartments with, until 1995, a stronger presence of these premises in the most exposed areas. This observation is also confirmed by interviews conducted in the field with decision-making actors, as this example shows:

> Then there is the plant which is exposed a little bit, the activity area too, but it is an activity area [...] (Interview with a community technician conducted in 2017, MobiClimEx project).

With the implementation of the Barnier law, municipalities are redirecting the construction of their business or industrial premises more frequently in areas of moderate risk. This figure highlights the differences in policy choices that mainly seek to spare individual homes.

2.4. Premises, contrasting situations

> The heterogeneity of medium- and long-term effects by type of exposure, history and risk culture of residents suggests that the overall effect is most likely to vary significantly depending on local circumstances and may explain differences in estimation results [MAU 15].

As we have said before, in 60 years, the population of the Gard has almost doubled to 733,201 inhabitants in 2013. The peri-urbanization process marked since 1970 around the major French conurbations, with in particular a tremendous expansion of individual housing, has not escaped this department, which nevertheless remains, on a French scale, a rather rural department. In the municipalities most subject to urban pressure, floodplains have been largely destroyed. This is why we wanted to show the differences in urbanization in flood-prone areas by taking into account the population size of the communes. We have opted for a division of the communes into six population classes:

– less than 400 inhabitants (112 in number);

– between 400 and less than 1,000 inhabitants (104 in number);

– between 1,000 and less than 2,000 inhabitants (64 in number, including Collias);

– between 2,000 and 10,000 inhabitants (65 in number including Le Vigan, Les Salles-du-Gardon, Remoulins, and Sommières);

– between 10,000 and 50,000 inhabitants (7 in number, including Alès and Bagnols-sur-Cèze);

– Nîmes with a population of more than 150,000.

Figure 2.5 shows the municipalities with a PPRI. It proposes for each of the groups its average trajectory of buildings in flood-prone areas by calculating for each year the proportion of buildings constructed in flood-prone and non-flood-prone areas. It highlights three pieces of information: first, over the past 60 years, a downward trend in urbanization in floodplains in all groups of municipalities except Nîmes; second, more frequent urbanization in floodplains in municipalities with more than 2,000 inhabitants due to greater urban pressure; and finally, the specific nature of the city of Nîmes.

"Local circumstances" go beyond strict soil management in a context of a permanent flood threat. The example of Nîmes cannot be mirrored from the situation of small municipalities where "there is no pressure to build in flood-prone areas, when there is room elsewhere"[26]. It is necessary to reiterate here the precautions to be taken so as not to draw overall conclusions for territories with heterogeneous trajectories. Population dynamics is the main marker [BOU 15] conditioning the contrasting exposure of territories to floods.

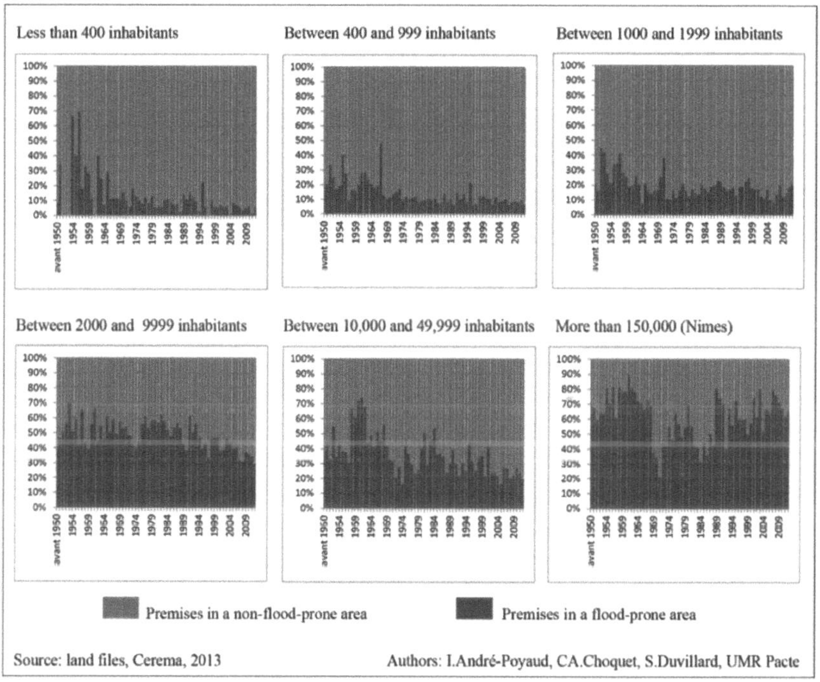

Figure 2.5. *Distribution of 2012 premises located in flood-prone and non-flood-prone areas according to their year of completion in municipalities with a PPRI. For a color version of this figure, see www.iste.co.uk/lutoff/hydrometeo2.zip*

2.4.1. *Typology of the municipalities of the Gard according to exposure to flood risk: summary in 2012*

The previous section allowed us to highlight different urbanization trajectories at the scale of groups of municipalities, where the demographic

26 Excerpt from an interview with a technician, as part of the ANR Mobiclimex, Gard 2017.

influence has exerted and still exerts a real pressure on the choices to urbanize in floodplains, even if, since 1995, the most exposed areas have been prohibited from construction. That said, the regulatory norm does not require the most exposed buildings to be systematically destroyed [HUT 15] and indeed the political choices that have followed one another over the years in the various municipalities of the Gard now lead municipalities to handle sometimes complex situations for their territory and their population in the event of an extreme hydrometeorological event.

We now propose to report on the spatial dimension of exposure to the risk of flooding by developing a typology of the municipalities of the Gard with regard to their stock of premises located in flood-prone and non-flood-prone areas. To do this, we combined two multivariate data analysis methods, first a simple correspondence factor analysis (CFA) and then a hierarchical bottom-up classification (HFA) (see Box 2.3 "Factor analysis and classification") on the first three factor axes of the FCA.

For each municipality, we identified the number of premises available in 2013 according to type (apartments, houses, and business or industrial premises) and their location (flood-prone or non-flood-prone areas). The six variables (number of houses in FPA, number of houses in NFPA, number of apartments in FPA, number of apartments in NFPA, number of business premises in FPA, and number of business premises in NFPA) played the role of active variables in determining the axes. Illustrative variables describing the population dynamics of municipalities and their exposure to flood risk were also added. These were the average annual population growth rates between the intercensal periods 1968–1982, 1982–1999, 1999–2006, and 2006–2014; the size of the municipality according to the population size; the proportion of its surface area in flood-prone areas; and the number of natural disaster orders for flooding recorded in the Gaspard database[27].

The factor analysis highlighted five hierarchical factorial axes. We have retained the first three axes, which retain almost all the information contained in the initial table, i.e. 98%. The first axis, which explains 62% of the total information, highlights a contrast between municipalities with a very high over-representation of individual housing in non-flood-prone areas and municipalities where collective housing and business or industrial premises

27 This database, developed by the French Ministry of Ecology and Sustainable Development, has been used since January 1, 1983, to identify the recognition of the state of "natural disaster" for all French municipalities.

are over-represented in flood-prone areas. The second factorial axis, which takes into account 26% of the total information, classifies the municipalities according to their distribution of houses in flood-prone areas. The third axis provides 9% of the total information. It pits municipalities for which apartments and business premises are more frequently located in non-flood-prone areas against municipalities where these types of premises are less so.

Factor analysis and classification methods are complementary. On large volumes of data, it is possible to carry out a factor analysis before a classification.

Factor analysis methods aim to reduce the number of variables in a table by proposing new factors that are hierarchical and independent of each other. The first factor highlights the strongest link between the observed variables/modalities and reflects the greatest direction of elongation of the scatter graph. The second factor identifies the second strongest link not taken into account by the first axis and then afterward. It is a question of retaining a factor subspace that preserves the strongest interrelationships and puts aside tenuous information.

In the case of our work, we implemented a factorial analysis of simple correspondences to operate a contingency table distributing in each commune (the rows of the table) the number of premises according to their nature and location (columns of the table). To interpret the selected factorial axes, it was necessary to define the elements responsible for the construction of an axis while keeping those having a strong contribution to the axis and good quality of representation on the observed axis. Then, we looked at the coordinates on the axis of the selected elements to locate them (negative or positive side). The factorial axes thus highlight opposition games between elements that are on either side of the axis and interrelationship effects between elements located on the same side of an axis. It is possible to add illustrative variables that help in the interpretation of the axes. However, these variables are not taken into account in the determination of the axis.

Classification methods make it possible to classify individuals into homogeneous groups. A classification is carried out on initial quantitative variables or from the factorial coordinates of individuals on factorial axes. The hierarchical bottom-up classification method (used here) consists of aggregating, in a bottom-up way and by successive iteration, a statistical individual at the center of the class closest to it. Initially, all statistical individuals form a class and at the end of the iteration, all individuals are brought together in a single class. To group the elements together, we used the Ward criterion (variance criterion) which involves at each step searching for a partition such that the internal variance of each group is as small as possible. Its successive groupings are displayed in a hierarchical tree or dendrogram. The objective is to cut the tree into a partition into x classes so that individuals who belong to the same class are as similar as possible and the classes are as different from each other as possible. The difficulty of this method lies in the choice of the partition among the proposed set and in the assignment of a name to each of the classes which must best reflect the salient characteristics. An average class profile is thus generated from the frequencies or averages of the active and illustrative variables. To interpret the classes, the modalities of the qualitative variables very strongly over- or under-represented and the means of the statistically significant quantitative variables were used.

Box 2.3. *Factor analysis of simple correspondence and hierarchical bottom-up classification*

At the end of this factorial analysis, we carried out a typology of these communes using the hierarchical bottom-up classification (HFA) method. The HFA was conducted on the coordinates of the communes on the first three factors and the illustrative variables. We proposed to retain a score in five classes. Figure 2.6 spatializes the results of this typology and Table 2.1 presents different statistical indicators associated with these groups of municipalities (number of natural disaster decrees, distribution by municipality size, and average annual growth rates over different periods).

	Class 1	Class 2	Class 3	Class 4	Class 5	Class 6	Gard
Number of communes							
	132	151	29	28	5	8	353
Number of municipalities by population class							
less than 400 hbt	70	38	2			2	112
400 to 1,000 hbt	38	57	5	4			104
1,000 to 2,000 hbt	16	30	9	7		2	64
2,000 to 10,000 hbt	8	26	8	17	4	2	65
10,000 to 50,000 hbt			5			2	7
More than 50,000 hbt					1		1
Population size in 2013							
Average	729	1,270	5,062	3,253	33,833	4,780	2,077
Minimum in 2013	12	55	242	540	2,618	216	12
Maximum in 2013	8,573	6,863	40,711	8,608	150,564	11,833	150,564

Average annual growth rate							
between 2013 and 2006	1.56	1.84	0.29	1.09	0.6	0.27	1.01
between 2006 and 1999	1.52	1.95	0.55	1.54	1.26	1	1.32
between 1999 and 1990	1.65	1.99	–0.37	1.38	0.49	0.99	0.9
between 1990 and 1982	2.15	2.76	–0.06	3.23	0.45	2.46	1.41
between 1982 and 1975	1.87	2.6	0	3.49	–0.5	2.46	1.01
between 1975 and 1968	0.38	1.07	–0.38	2.15	0.42	1.75	0.49
Natural disaster shutdowns							
Number of events	621	1,050	225	197	45	64	2,202
Average	4.7	7	7.8	7	9	8	6.2
minimum	1	2	4	2	4	3	1
maximum	12	13	14	12	16	14	16

Table 2.1. *Populations and natural disaster orders of the six profiles of municipalities in the Gard*

The first group, in white on the map, includes the 140 municipalities that did not yet have a PPRI approved in 2016. The totality of the buildings in these municipalities is *de facto* recorded in a non-prone area. Nevertheless, even without PPRIs, the populations and buildings of this group of municipalities may be exposed to the risk of flooding. Indeed, even if these municipalities are on average less exposed to this risk than those of the other classes of the typology, they have all already suffered damage with at least one declaration of natural disaster decree due to flooding, runoff, or mudflow since 1983. Indeed, they declared an average of 4.7 natural disaster orders for flooding compared to 6.2 for the whole of the Gard. It should be noted that the communes of Génolhac and Savignargues have been the subject of at least 10 natural disaster decrees of this type.

The second group, represented in green on the map, consists of 151 municipalities characterized by poorly built building in flood-prone areas; 91% of the premises are located in non-flood-prone areas compared to 64% for the whole of the Gard. These municipalities are more akin to a profile of rural municipalities – 63% of them have less than 1,000 inhabitants and 20% have populations of between 1,000 and 2,000 inhabitants where individual housing is very largely over-represented. 71% of the premises are detached houses built in non-flood-prone areas, i.e. 30% points more than for the Gard as a whole. On average, these municipalities have recorded nearly seven natural disasters due to flooding, which places them just above the average in the Gard.

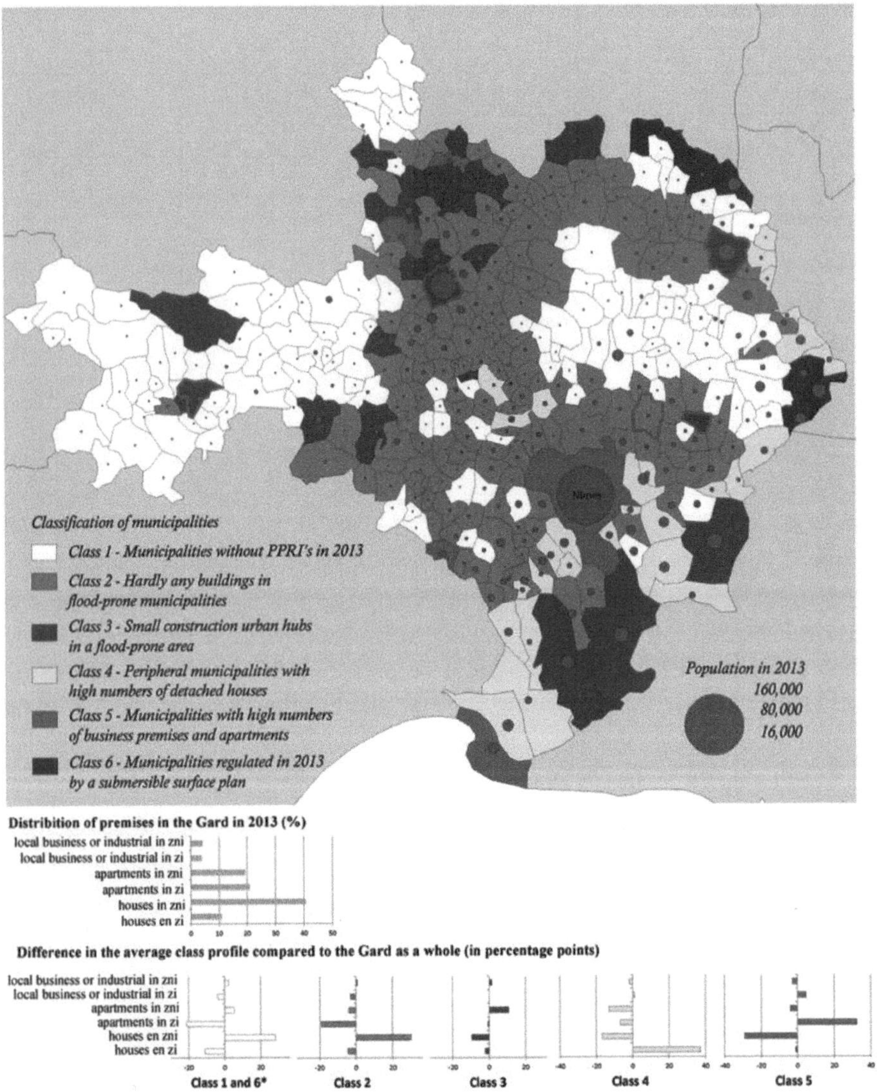

* : The distribution of premises in flood-prone areas could not be calculated in municipalities regulated by a PSS.

Sources: IGN - Géofla, INSEE - 2013 population census and Cerema - Land files 2013
Director: Isabelle André-Poyaud, Charles-Antoine Choquet, Sylvie Duvillard, UMR Pacte, 2017

Figure 2.6. *Typology of municipalities according to the floodability of their buildings in 2013 with regard to their population. For a color version of this figure, see www.iste.co.uk/lutoff/hydrometeo2.zip*

Collias, one of the seven municipalities studied[28], falls into this category. Today, although this municipality has buildings that are not very exposed to the risk of flooding, it was strongly affected by the floods of September 2002. The data available in the 2013 land files database does not allow us to report on this episode of flooding that devastated one of these neighborhoods built in the major bed of the Gard. This river, rising in 24 hours a total of about 20 m, submerged 95 houses, five of which were completely destroyed. Faced with the seriousness of the damage, the mayor of the time informed the prefecture, which prohibited the renovation of the most exposed damaged premises. Following this tragic event, 48 houses were razed to the ground and their inhabitants relocated to new housing estates built on the heights of Collias.

The third group, in purple on the map, includes 29 municipalities, including small urban centers. Fourteen municipalities have more than 2,000 inhabitants. In this class, the share of premises of all categories and located in flood-prone areas represents 33% of the premises. This class is characterized by an over-representation of collective housing premises in non-flood-prone areas and proportions close to the average profile for other types of premises. This distribution is explained by a relatively small footprint in the floodplain. For 80% of these municipalities, i.e. 23 municipalities, the floodplain envelope is less than 1/4 of the municipal surface area. This over-representation of residential premises built in non-flood-prone areas remains valid, including for the four municipalities (Saint Gilles, Remoulins, Beaucaire, and Valvert) most exposed to flooding with more than half of their territory defined as flood-prone areas. The municipalities of Alès and Bagnols-sur-Cèze were included in this class of the CAH (Classification ascendante hiérarchique - Hierarchical ascending classification), even if their profile differs by a greater over-representation of collective housing in flood-prone areas (respectively, 34% and 27% of apartment type premises located in flood-prone areas compared to 20% for the whole of this class). With the exception of Alès and Bagnols-sur-Cèze, buildings, whatever their use, are on average less exposed than in the other municipalities in the Gard region that are subject to a PPRI.

The fourth group, in yellow, includes 28 municipalities located on the outskirts of Nîmes and near the main roads of the department. Due to their location, these municipalities have attracted new populations to them, most of whom were looking for detached housing during the five intercensal periods between 1968–1982, 1982–1989, 1999–2006, 1996–2006, and

28 As part of the ANR Mobiclimex.

2006–2014, even if this attraction occurred mainly between 1968 and 1990. Over the years, urbanization has affected the floodplains of these municipalities. In 2013, two-thirds of the premises were located in flood-prone areas, 70% of which were detached homes. This class was distinguished from the others by its very high over-representation of individual housing type premises in flood-prone areas.

The fifth group, in red on the map, consists of five municipalities with between 2,000 and 10,000 inhabitants (Les Salles-du-Gardon, Sommières, Le Grau-du-Roi, and Bessèges) and the Gard capital, Nîmes. These municipalities all have a high exposure to the risk of flooding, as shown by their number of natural disaster decrees (16 for Nîmes, 10 for Sommières, and 9 for Salles-du-Gardon). In terms of urbanization, in 2013, 72% of the premises in this class were built in flood-prone areas, with a very high over-representation of collective housing and business or industrial premises. In the city of Sommières, 9 out of 10 apartments are built in flood-prone areas. These apartments, most of which are located in the old town center, are located on at least the first floor of the buildings. Since the 2002 flood, the ground floors of the medieval center have almost all been occupied by business premises, bringing the proportion of business or industrial premises in flood-prone areas to 86%. Our interlocutors surreptitiously led us to consider the specificity of each situation, recalling for example that for the 1958 flood in the Gard, "the upstream dams, in the collective imagination it was the response to the floods, therefore the whole development of the right bank (flood-prone) took place in the 1960s," then "2002 recalled [...] the flood"[29]. After 2002, there was still construction in the floodplains in the old center with specific rules in dense urban areas, which allowed for requalification operations for urban renewal and collective housing[30]. The impact of the event is interpretable regardless of the year in which it took place (1958/1988/2002) and combined with the implementation of a standard (1995), a virtuous dynamic is observed.

The sixth group (in gray) consists of municipalities included in submersible surface plans. These are the municipalities of Pont-Saint-Esprit, Aiguèze, Le Garn, Saint-Julien-de-Peyrolas, and Saint-Paulet-de-Caisson for the PSS Ardèche, approved on March 27, 1959, and still in force in 2013,

29 Excerpt from an interview with a technician, as part of the ANR Mobiclimex, Gard 2017.

30 Above the PHE ratings, hence the building permits and the presence of a new building in the DGfip base, verbatim an elected representative and a technician from two communes outside Nîmes.

and Pujaut, Les Angles, and Villeneuve-les-Avignon for the PSS Rhône Amont approved on August 6, 1982. As we do not have the spatial hold of these PSSs for these municipalities and cannot calculate the distribution of their premises in flood-prone or non-flood-prone areas, we preferred to isolate these municipalities without PPRIs (class 1), as their recognized flood risk could be considered higher. It also appears that these municipalities recorded an average of eight natural disaster orders for floods compared to an average of 4.7 for municipalities without a PPRI.

This typology makes it possible to highlight several logics of urbanization in the communes of the Gard. These strategies take into account the physical constraints of their territory, their industrial history and the territorial dynamics in which they are embedded. Whenever possible, the municipalities built preferably in non-flood-prone areas. This is often the case in small rural municipalities. Municipalities that play a central role and are subject to high exposure to the risk of flooding have opted more for collective housing and local business or industrial buildings in flood-prone areas. The municipalities on the outskirts of Nîmes, caught up in a process of urban sprawl and to satisfy the residential preferences of the French seeking homeownership in a detached-style home [DAM 17], have largely opened their plots to an individual housing type for construction, including those located in floodplains. In this category, there is even greater exposure for individual open-water habitats that have developed significantly in recent years.

2.4.2. *Time agreement between recurrence of events and reduction of individual housing in flood-prone areas: the example of Nîmes*

Given its role as the leading urban center, Nîmes has, and continues to, urbanize in flood-prone areas despite its high exposure to flooding. Built on the banks of the Vistre, Nîmes is crisscrossed by numerous *cadereaux*[31] that overflow very frequently in flood-like situations. In contrast to the whole of the Gard, in this city, the proportion of buildings in flood-prone areas is just as high today as it was in the 1950s and 1960s, in a context of strong attractiveness. In Nîmes (Figure 2.7), 75% of the apartments (n apartments FPA = 47,115) and 85% of the business or industrial premises (n business premises FPA = 9,346) are located in flood-prone areas. However, the

31 A local French term that refers to streams or torrents, usually dry, draining water surrounding scrubland and the crisscrossing of the city of Nîmes.

occurrence of events at increasingly shorter time intervals seems to have a lasting effect on house construction in flood-prone areas. However, in 2012, there were 32% of houses in flood-prone areas (n houses FPA = 9,595). The hypothesis that the recurrence of events close in time increases the ability of actors to metabolize risk [MAU 15], [PRY 11] by reducing urbanization is supported (Figure 2.8).

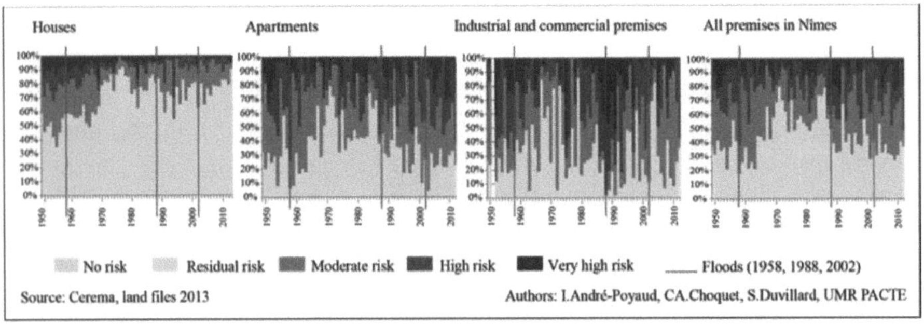

Figure 2.7. *Distribution of types of premises by year of completion and risk level in Nîmes. For a color version of this figure, see www.iste.co.uk/lutoff/hydrometeo2.zip*

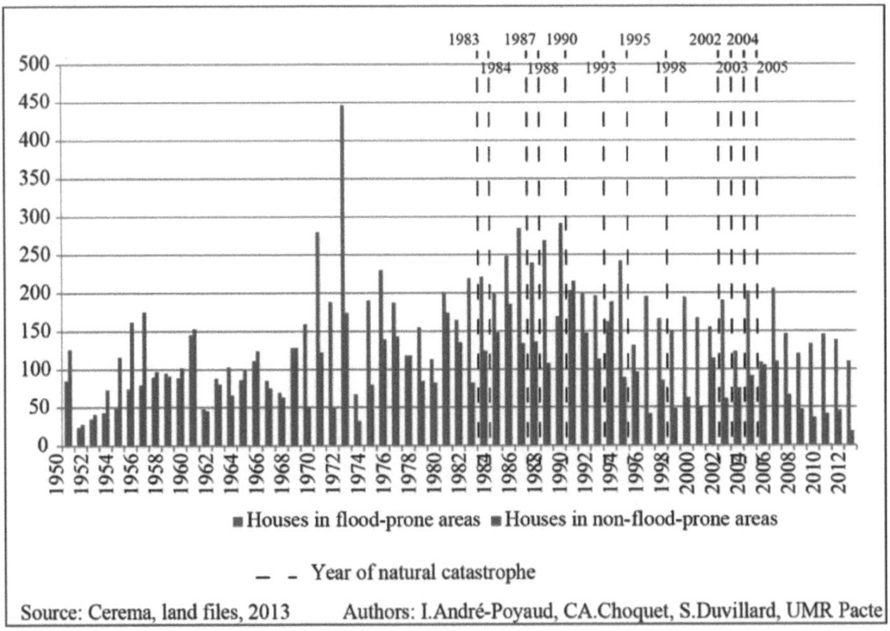

Figure 2.8. *Number of houses built in Nîmes in 2012 per year of completion and location. For a color version of this figure, see www.iste.co.uk/lutoff/hydrometeo2.zip*

Since 1995, the number of houses built in flood-prone areas has been lower than that in non-flood-prone areas. The combined effects of 1995 and the recurrence of events reversed the trend and the continuation of the urbanization dynamics. Further urbanization is taking place "high up" (Figure 2.7). However, the houses built subsequently were those before the revision of the PPRI in 2011: compared to the 2008 revision, the gutter houses located to the north of the city were not in a floodplain; after 2011, the PPRI lost its "size"[32] in the city center, to encompass more in detail the banks of the streams flowing down the northern hills of the city, where urbanization is dense but in detached home areas (Figure 2.9).

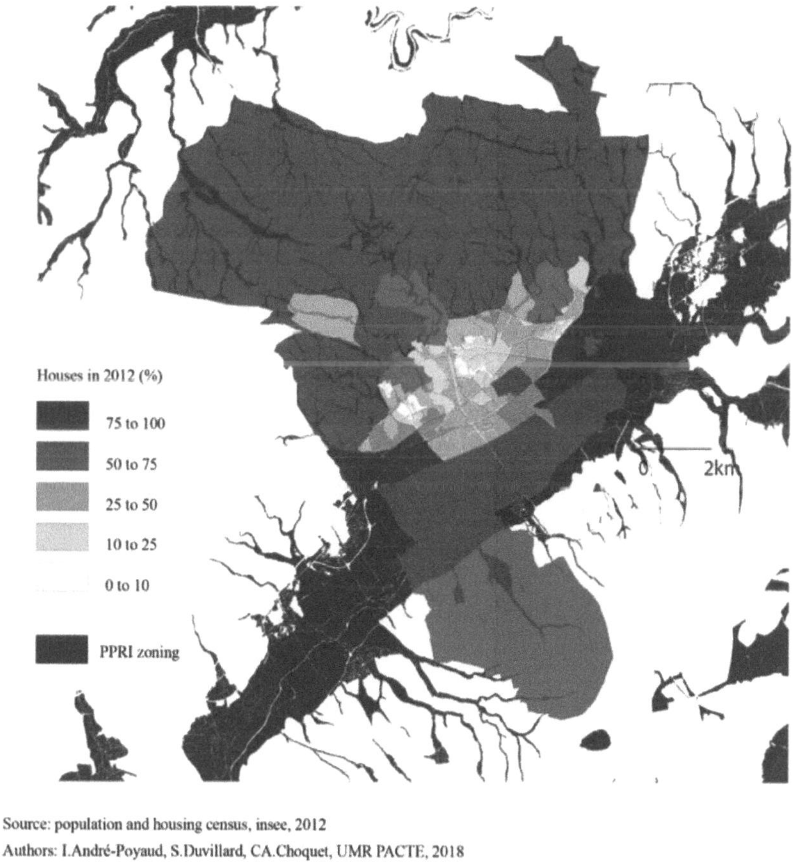

Figure 2.9. *Percentage of houses by Iris in Nîmes in 2012 and PPRI scope*

32 Narrowing in the old city center.

The Nîmes example shows the compromises to pursue the city's development: changing the zoning ZACs[33], "the ZAC we took out of the PPRI [...] we took it out slowly"[34]. The introduction of a new class "between the very high and the moderate" should make it possible to densify certain areas in relation to the transport axes into exclusive sites, "there were no 'subtleties' involved"[35].

In other localities, the PPRI was integrated into the PLU in advance after 2002: "filling the city and densifying explains the continuation of construction in floodplains by renovating old premises. It is possible to make extensions on the ground floor for economic activity, the housing is on the upper floors, [...] so we reduced vulnerability"[36].

Interviews with local stakeholders suggest that caution should be exercised in analyzing the data: the construction of residential premises in flood-prone areas sometimes masks the "cut-to-sale" market and the return to the market of old apartments located in the historic center. They are recorded as new builds, which explains the continued urbanization in high-risk areas. "Everything that was flood-prone was built; the 1980s were marked by densification in floodplains, followed by a decline between 1995 and 1998; if there was no restart in 2002 in floodplains, it was because everything was already built"[37].

33 *Zone d'aménagement concerté* (concerted development zone): concerted development zones are those within which a public authority or public institution whose purpose is to intervene to carry out or have carried out the development and provided facilities for the land, in particular those that this authority or institution has acquired or will acquire with a view to subsequently transferring or granting them to public or private users. The scope and program of the concerted development zone are approved by deliberation of the municipal council or the deliberative body of the public establishment for intermunicipal cooperation. http://caue.com/le-conseil-aux-collectivites/memento/zac/.

34 Excerpt from an interview conducted with a technician, as part of the ANR Mobiclimex, Gard 2017.

35 The Gazette des communes concerning the PPRI de Nîmes, http://www.lagazettedes communes.com/103335/risque-inondation-le-ppri-de-nimes-reoriente-le-developpement-de-la-vil le, published on March 1, 2012, consulted on February 27, 2018.

36 Maintenance technician, Gard 2017.

37 Stakeholder interviews, Gard elected representative 2017.

2.5. Conclusion

From a methodological point of view, this study shows the interest of using the source of land files to reconstruct the rates of urbanization of a territory over a long period of time, even if these observations are based on an existing stock of buildings and premises which does not make it possible to account for the destruction or changes in the nature of the destinations of the properties.

The interpretation of these observations according to the prism of flood risk shows three periods: before 1975, when constructions in flood-prone areas were more frequent than in non-flood-prone areas; between 1974 and 1995, with sustained construction activity in both types of areas; and after 1995, when the influence of the Barnier law led to the abandonment of flood-prone areas without, however, being definitively abandoned. In addition, the occurrence of extreme hydrometeorological events appears to produce short-term breaks in the observed rates. These declines are more likely to be seen within 2–3 years of the phenomenon.

The results show differences within the Gard territory. Among the municipalities exposed to this risk, those subject to land pressure remain more concerned by urbanization in flood-prone areas. Differences also exist in terms of the type of premises built in flood-prone areas. Even if today, decision-makers prefer the construction of business premises and buildings, the fact remains that municipalities located along the main roads have massively built individual housing which is still highly exposed today.

The trade-offs and compromises that decision-makers seem to be reaching are difficult to overcome from the sometimes contradictory constraints and injunctions (less urbanization in the industrial zone but densification to save land). It is up to them to place the cursor on the acceptability of the risk according to the exposure of uses and the recurrence of events over time.

2.6. References

[BON 16] BONNET F. (ed), *Atout risques; des territoires exposés se réinventent*, Territoires en projet, Parenthèses, Paris, 2016.

[BOU 15] BOULIER J., DUBOS-PAILLARD E., PAVARD A., "Évolution de la vulnérabilité urbaine face au risque de crue rapide dans le Gard", in MATTEI M.-F., PUMAIN D. (eds), *Données urbaines 7*, Economica, Paris, 2015.

[BOU 18] BOULIER J., DUBOS-PAILLARD E., PAVARD A., "The difficulties in local implementation of risk prevention policies", in LUTOFF C., DURAND S. (eds), *Mobility in the Face of Extreme Hydrometeorological Events 1: Defining the Relevant Scales of Analysis*, ISTE Press Ltd, London, UK, pp. 151–185, 2018.

[DAM 17] DAMON J., "Les Français et l'habitat individuel: préférences révélées et déclarées", SociologieS [Online], Files, Où en est le pavillonnaire?, February 21, 2017, accessed October 2, 2019, available at: http://journals.openedition.org/sociologies/5886 .

[DOU 11] DOUVINET J., DEFOSSEZ S., ANSELLE A. *et al.*, "Les maires face aux plans de prévention du risque inondation (PPRi)", *L'Espace Géographique*, vol. 40, pp. 31–46, 2011.

[HUB 03] HUBERT G., CAPBLANCQ J., BARROCA B., "L'influence des inondations et des documents réglementaires sur le marché foncier en zone inondable", *Annales des Ponts et Chaussée*, no. 105, p. 32, 2003.

[HUT 15] HUTEAU C., "Gestion des risques littoraux: adaptation ou déplacement de populations?", in LARONDE-CLERAC C., MAZEAUD A., MICHELOT A., *Les risques naturels en zones côtières; Xynthia: enjeux politiques, questionnements juridiques*, Presses Universitaires de Rennes, Rennes, France, pp. 75–89, 2015.

[KOU 10] KOUSKY C., "Learning from extreme events: risk perceptions after the flood", *Land Economics*, vol. 86, no. 3, pp. 395–422, by the Board of Regents of the University of Wisconsin System, 2010.

[LAG 06] LAGANIER R., *Territoires, inondation et figures du risque; la prévention au prisme de l'évaluation*, L'Harmattan, Paris, 2006.

[LAR 15] LARONDE-CLERAC C., MAZEAUD A., MICHELOT A., *Les risques naturels en zones côtières; Xynthia: enjeux politiques, questionnements juridiques*, Presses Universitaires de Rennes, Rennes, France, p. 205, 2015.

[MAU 15] MAUROUX A., "Exposition aux risques catastrophiques, politiques de prévention et marchés de l'immobilier en France – État de la connaissance en économie", *Études & Documents*, Commissariat général au développement durable, no. 134, 2015.

[MAZ 15] MAZEAUD A., "Quand la catastrophe fait de l'urbanisation en zones côtières une prise de risque inacceptable; logique d'imputation des responsabilités et dépolitisation de la gestion des risques naturels", in LARONDE-CLERAC C., MAZEAUD A., MICHELOT A., *Les risques naturels en zones côtières; Xynthia:enjeux politiques, questionnements juridiques*, Presses Universitaires de Rennes, Rennes, France, pp. 121–139, 2015.

[MES 06] MESCHIMET de RICHEMOND N., SCARWELLH J., FORT M. *et al.*, "Restituer le territoire au risque", in LAGANIER RICHARD (ed), *Territoires, inondation et figures du risque; la prévention au prisme de l'évaluation*, L'Harmattan, Villeneuve d'Ascq, pp. 177–215, 2006.

[MOU 14] MOULINE., DEROUBAIX J.-F., HUBERT G., "La constructibilité des zones inondables ou la négociation territoriale de la règle PPRI", *Géocarrefour* [Online], vol. 88, no. 3, published online June 12, 2014.

[PRY 11] PRYCE G., CHEN Y., GALSTER G., "The impact of floods on house prices: an imperfect information approach with Myopia and Amnesia", *Housing Studies*, vol. 26, no. 2, pp. 259–279, 2011.

[RUI 08] RUIN I., CREUTIN J.D., ANQUETIN S. *et al.*, "Human exposure to flash floods – relation between flood parameters and human vulnerability during a storm of September 2002 in Southern France", *Journal of Hydrology*, vol. 361, nos 1–2, pp. 199–213, 2008.

[SCA 14] SCARWELL H.J., *Urbanisme et inondation, outils de réconciliation et de valorisation*, Septentrion, Presses universitaires, Paris, 2014.

[SER 00] SERRAT P., CALVET M., "L'urbanisation d'une plaine inondable: le risque oublié? La Salanque (Pyrénées-Orientales) face à la crue des 12 et 13 novembre 1999 [The urbanisation of a plain liable to flooding: a forgotten risk? The Sallanque (Pyrénées-Orientales)]", *Géocarrefour*, vol. 75 no. 3, pp. 209–220, 2000.

[TRO 04] TROY A., ROMM J., "Accessing the price of flood hazard disclosure under the California natural hazard disclosure law (AB 1195)", *Journal of Environmental Planning and Management*, vol. 47, no. 1, pp. 137–162, 2004.

[VAL 09] VALY J., "L'implantation urbaine en zone inondable: exemple de deux communes d'Ile-et-Vilaine", *M@ppemonde*, vol. 96, no. 4, 2009, available at: https://mappemonde-archive.mgm.fr/num24/articles/art09404.pdf.

3

Factors Influencing Residential Choices in Flood-Prone Areas: From Principles to Actors' Actual Practices

3.1. Introduction

Since risk is defined as the encounter between a risk and an exposure, the risk incurred by individuals is inseparable from their residential choices. The choice of home location determines – at least in part – the exposure of residents to risk in their place of residence (non-mobile exposure). In support of a case study conducted in southeastern France – particularly exposed to flash floods – this chapter explores the factors influencing these residential choices in flood-prone areas.

This question regarding the impact of the flood on residential choices was mainly addressed from an economic perspective. According to the dominant hedonic property theory [ROS 74], the risks of imposing a form of servitude attached to real estate are reflected in house prices [MAU 15a]. While there is a great deal of work to be done in other national contexts, particularly in the United States, the impact of natural hazards on French real estate markets remains poorly known; the available work is limited to localized case studies based on transaction prices, the results of which do not allow a decision to be taken on the subject. Some post-disaster case studies trace depreciative effects on sale prices in the aftermath of disasters; [DER 06], [HUB 03], [LON 01] when others show that this effect does not last over time or that the amenities offered – particularly proximity to the river – or the

Chapter written by Séverine DURAND, Céline LUTOFF and Sylvie DUVILLARD.

practicality of the location more than offsetting the negative effect of exposure to risk [DUR 15], [HUB 03], [LON 01], [MAU 15a]. Properties located in high-risk areas are sometimes even paradoxically valued by the insurance of the impossibility of new construction [HUB 03]. What is shared, however, is the great complexity of these links and the difficulty in clearly demonstrating an effect and its maintenance over time [MAU 15a]. Some authors defend their indeterminacy – in terms of variability in terms of disasters, local history, and spatial dynamics of short-term effects [MON 94]. Our work on living in floodplains [DUR 14] and then on residential trajectories in floodplains [DUR 18] illustrated how multiple factors determine residential choices, and how the question of risk can be reduced by a set of other criteria related to the practicality of the places [DUR 14] and the attachments that link residents to their living environment and each other [DUR 18]. However, the explanations available in the literature, mainly in microeconomics, on this lack of persistence over time of the influence of flooding on the land market or even of a lack of spatial variability of property prices that does not reflect differences in risk exposure, point rather to biases in perception due to imperfect and incomplete information or unconscious risks [MAU 15a]. In the specific cases of "extreme" or "catastrophic" risks – as in the case of flash flood phenomena that motivates this book – the importance of personal experience on decisions [TAL 07] is stressed in the face of complex information in situations of uncertainty, and where decision making depends not only on the characteristics of the decision but also on the context in which it is taken [MAU 15b]. The explanation of availability[1] bias is the most defended [CGD 14]: people who have already experienced a disaster would be the most likely to protect themselves from it, and therefore, to consider this question, particularly in their residential choices.

In this chapter, we propose to feed the reflections from a sociological point of view by studying the two elements likely to influence residential choices: the law and the principle of information for buyers and tenants to risk exposure and prices. For this reason, we are interested in study areas where the question of flood risk is present, in a department that regularly makes the headlines for dramatic floods[2]. Through a qualitative survey of

1 The more a random event occurs historically, the less easily the individual remembers it and the less likely the occurrence of risk will seem [TVE 73].

2 The Gard is a department highly exposed to floods. More than 500 floods have been recorded since the middle of the 13th century. The national rainfall record was set in 1907 with 950 mm in 24 hours. The vulnerability of the Gard to flooding is increasing. In 2013,

residents and professionals from three municipalities differently exposed in this territory, it is, first of all, a question of the practical reality of applying the principle of law: is the information provided in texts well transmitted? The second step is to question the effects of cases where information is transmitted, and thus more broadly to question the determining nature of the flood on residential choices. Finally, we will question the influence of the land market on these choices. To do this, we propose to cross-reference the analyses from the point of view of real estate professionals on the conditions of mobility in areas exposed to the analysis of the residential trajectories of the inhabitants. We will first present our working methodology (section 3.2.1) as well as the specificities of each case study (section 3.2.3) that will support the argument. We will then present our analysis from three thematic perspectives. We will focus on residential choices by questioning the information conditions prevailing for this choice (section 3.1) and more broadly on the determining nature of the risk (section 3.2) and the price (section 3.3) following this choice.

3.2. Socio-ethnographic investigation on living practices and narratives by land professionals

3.2.1. *Methodology*

The analysis presented in this chapter is based on a qualitative study carried out at a micro level among inhabitants on the one hand (a) and land professionals on the other hand (b), which provides information on the individual motivations of residential mobility and land mobility dynamics observed at the meso level (see Chapter 2 of this book on these points). This study was part of the research project funded by the French National Research Agency: the MobiCLIMeX[3] project.

259,000 people lived in flood-prone areas, representing 35% of the Gard population. 21% of the Gard's territory is in a floodplain, its urbanized surface represents 9% of the territory, a quarter (25.6%) of which is in a floodplain. During the floods of September 2002, 299 municipalities out of 354 were affected, 22 deaths were reported, and the damage amounted to €830 million. The 2003 floods affected 37 municipalities and caused €300 million of damage, and the 2005 floods affected 86 municipalities and caused €26.65 million of damage. Sources: http://www.noe.gard.fr.

3 This Mobiclimex project was financed by the 2012 ANR Société & Environnement program and is part of the international HYMEX (HYdrological cycle in the Mediterranean EXperiment) program on the change of water cycle in the Mediterranean basin in a climate change situation. This last project focuses on the change in exposure to flash floods in south-

– With regard to individual practices, a multi-site socio-ethnographic investigation of residential trajectories in floodplains[4] has aimed to reintegrate the study of residential choices into the broader framework of "life trajectories" that involve more globally different dimensions of social beings' involvement [GRA 10]. The survey aimed to replace the question of the risk of flooding in the daily lives of inhabitants, i.e. as one of the dimensions of a dwelling considered in its entirety. The interest in observing the ordinary practices of living in a risky environment is that these practices could teach us about whether or not the risky dimension of the environment, and its influence on residential choice, should be taken into account. The survey was planned over a long period of time (24 months), taking care to multiply the temporalities of the survey (observation practice and conduct of individual interviews) in the ordinary life of the spaces. Moving away from the event leads to considering the question of risk – not simply as it, but living within it on a daily basis, i.e. studying the "risk public," those "living with risk" [DUR 14], [NOV 08].

The protocol was characterized by a first phase of observation in the field (i), monitoring contemporary local concerns (ii), analysis of the sociological, economic, and political elements framing the situation (iii), and conducting semi-directive interviews (iv). 27 semi-directive interviews were conducted with residents (11 men and 10 women) or couples (6 interviews), for a total of 33 inhabitants. The interviews aimed to consider the daily experiences of inhabitants, to pay attention to the meaning given by the actors to their own behaviors and experiences. Without focusing solely on the issue of flooding, the interviews were framed around the person's personal history in relation to their habitat. From an ethno-sociological perspective, the interviews have aimed to orient the inhabitants' stories toward the form of narratives of practices. To do this, the method of life stories, here focusing on the narratives of living practices, was chosen. Reducing the observation field to a particular type of process (i.e. residential trajectories) and context (i.e. exposure of the territory to floods) allowed for a generalization of the overall formation of biographical trajectories would make impossible. This form of empirical data collection made it possible to monitor the formation of residential trajectories: it made it possible to understand the mechanisms and processes by which individuals have come to find themselves in a given

---

eastern France since the 1960s and more specifically in the Gard department, which is regularly subjected to this type of event.

4 The reader may refer to Appendix 2 of Volume 1 for more details on the survey methodology.

situation (in this case living on a floodplain) and how they managed this situation or tried to escape it [BER 05]. The sampling ensured a certain balance of population categories: we made sure to vary the traditional sociological categories (sex, age, and type of activity) of the respondents, but also the categories that could possibly influence the type of narrative sought, i.e. in this case: the location [on or away from the floodplain according to the regulatory mapping of the flood risk prevention plan (PPRI)], occupation status (tenant vs. owner), as well as the type of housing (town house, detached house with garden, apartment; first floor versus on the second floor). The corpus of narratives about living practices collected was the subject of a thematic analysis integrated with other materials collected through observation and documentary study. The comparison of biographical paths[5] allowed us to reveal recurrences, similar logics of action. It was through this comparison work that the initial or elaborated hypotheses during the comparative work could be clarified, confirmed, and took on a strictly sociological form [BER 05]. Replacing these analytical elements in the literature on residential trajectories entered more generally, and then made it possible to formulate hypotheses on residential trajectories in floodplains based on this case study.

– With regard to the qualitative study of local land dynamics, semi-directive interviews were conducted with local experts. These interviews were conducted in two stages. Exploratory interviews were conducted at the beginning of the parallel quantitative study on the pace of urbanization (presented in Chapter 2) that informs and supports the results presented here. Further interviews were conducted at the end of this study to ask actors about the results obtained. 18 interviews were conducted with experts in the municipalities concerned by the study: lawyers (3), risk managers (4), insurers (2), real estate agents, advisors or former municipal councilors (4), agents of local urban planning agencies or urban planning consultancy agencies (3), and flood risk managers (2). These interviews were the subject of a thematic analysis.

3.2.2. Areas of study

The chapter's argument is based particularly on the case studies of the cities in which the socio-ethnographic approach has been deployed, three

5 An example of a representation of residential trajectories (residential time line) is given in Appendix 4 of Volume 1.

cities exposed to the risk of flooding in the MobiCLIMeX[6] project study area. These cities were selected for their contrasts in terms of size, economic dynamics, and risk exposure. Interviews with professionals were conducted in each of these cities alongside the residents' survey. The cities were Sommières, Collias, and Le Vigan.

Collias is a village of 1,112 inhabitants[7], located near the famous Pont du Gard, in the east of the department. A small rural village historically known for agriculture and tourism, Collias is close to major traffic routes and is becoming increasingly attractive as urban centers near Nîmes and Avignon become more suburban. During the 2002 flood that affected almost the entire department, several houses were washed away and many residents had to be relocated. About a hundred homes were heavily impacted by the floods, some completely washed away by the river. Faced with the extent of the damage, the State (with the support of local authorities) used the Barnier fund[8] to expropriate 90 homes (48 principal and 42 secondary). The inhabitants concerned were offered the sale of their home, at a negotiated price, of a building plot on a non-flood-prone part of the municipality. It is this specificity that makes this area interesting for the survey by proposing a sample of families that have been relocated.

Sommières is a city of 4,644[9] inhabitants. This medieval city has the particularity of having been built on the very bed of the river. With a very rich history, marked by the manufacture and trade of silk, Sommières has enjoyed a constant attractiveness for centuries. Despite the regular floods that hit, it remains very attractive today because of its tourist interest and also because of its proximity to the main roads to reach the two urban centers near Montpellier and Nîmes. This city is interesting because it cannot be more protected against floods: there will always be floods.

6 See footnote 4.

7 INSEE 2014 population census.

8 The *fonds de prevention des risques majeurs* (the French major natural hazards prevention fund), FPRNM, (also known as the *fonds Barnier*, "Barnier Fund") was created by Act No. 95–101 of February 2, 1995 on strengthening environmental protection. This fund was then intended to finance compensation for the expropriation of property exposed to a major natural hazard. Its use was then extended to other categories of expenditure. The current scope of measures eligible for subsidies from the fund is defined in Article L561-3 of the Environmental Code. These measures and their application in this specific case are detailed below in the body of the text. The Barnier fund is funded by a 12% levy on the "natural disasters" premium for home and car insurance policies. (Sources: www.eure.gouv.fr).

9 INSEE 2014 population census.

Le Vigan is a town of 3,933[10] inhabitants located in a rural valley in the northwest of the Gard. A city in economic decline today, it has shone for centuries. Like many cities in the Cévennes, it owes a large part of its prosperity to the hosiery industry (silk spinning and knitting of stockings). The great wealth of traders came from the work of wool and sericulture, which made it possible to knit silk stockings for export throughout Europe. The city was the administrative capital of the Western Cévennes, important because of its large fairs and its role as a stopping place at the foot of the Cévennes. The textile industry flourished in the 20th Century. More recently, several company relocations have affected the local labor market. This city is interesting because of its difficult economic situation. Unlike the other cities in our sample, census data show that the population is decreasing substantially here.

3.3. The buyer's informed choice: no uninformed buyers

Legally, no one can settle in a risk area, whether as a tenant or owner, without being aware of the "natural" or technological risks to which the property is exposed. Since 2006[11], it has been mandatory, as part of the *Information Acquéreur Locataire*, IAL (tenant purchaser information), to provide information on the technological and natural hazards, affecting a property in the exposed municipalities. All sellers or lessors who are natural or legal persons governed by public or private law – including local authorities, the State, or their public establishments – are concerned by this obligation to provide information. This information takes the form of a risk statement to be drawn up less than six months before the date of close of sale, the contract of sale, or the written lease to which it will be attached. All types of real estate, built or unbuilt, are concerned, regardless of their destination. Thus, for 15 years now, people settling in floodplains have been theoretically informed of the risks to which they are exposed.

The IAL is an element of a risk management policy that experienced a "preventive turnaround" in the early 2000s [CRE 18], [DUR 16][12]. For a

10 INSEE 2014 population census.

11 According to Article 77 of Law 2003-699, implemented in January 2006.

12 Flood management consists of measures that are classified by flood specialists according to two main areas: protection and prevention, whose definitions are often confused by their widespread use. We consider here that protection includes measures aimed at controling phenomena and their consequences as much as possible when preventive measures,

long time, protection policies based on the technical control of risk by experts or public authorities largely predominated in the West. These have dramatically revealed their limits – notably through the series of natural and technological disasters that marked the end of the 20th Century and the beginning of the 21st Century in Europe. As a result, overall risk management strategies have been oriented toward more prevention by moving from flood control to promoting resilience to floods. In France, it is the "Risk Law" of July 30, 2003, also known as the "Bachelot Law," relating to the prevention of technological and natural hazards and the repair of damage, that has reflected this change of orientation in the texts. It is structured around five axes, the third of which is information on risks at the source[13]. This change in risk management policies is based on the premise that the effectiveness of a risk management policy depends on the knowledge of the risks by the populations concerned. This is based on the assumption that informed people are more likely to take preventive measures and thus react better when an event occurs [DUR 14].

According to the real estate professionals we met as part of our study, in practice, this principle of an IAL is respected. This fact seems to be obvious to the actors we met. Because their responsibility was engaged – as was the case with the lawyer – their respectability was also engaged, particularly for real estate agents covering a local market where non-knowledge dominates. According to the professionals we met, people involved in a purchasing process in particular know beforehand, most of the time. According to them, in terms of construction in particular, the majority of building permits are issued by people already residing in the municipality.

> They know exactly what they're buying these people. They're knowledgeable people, they already live there. Unfortunately, we do not sell much to people outside the area (a lawyer practicing and living in one of the study municipalities).

---

underpinned by a form of flood acceptance, invite people to deal with them, and thus prepare for them. Protection essentially includes facilities (dikes and dams) built to control the extent of flooding. Prevention includes measures put in place to mitigate the extent of the damage, institutionalized short-term prediction systems aimed at organizing vigilance and warning, as well as information systems on risks to residents. We refer the reader to Chapter 5 of Volume 1 for details on changes to French legislation on flood risk.

13 The other 4 principles are: (1) Strengthening information and consultation on major risks. (2) Controlling urbanization in high-risk areas. (3) Improving the conditions for compensating victims. (4) The development of awareness, recall, and appropriation of risk.

According to the lawyers we met, this effect remains limited. This information comes at a time when people are already engaged in a purchase or rental process (at the time the compromise is signed at the lawyer's office) and among a large amount of other mandatory information and diagnoses (energy performance, lead and asbestos exposure, etc.).

> We had deeds that were four or five pages long 40 years ago (...) and today, on average, deeds are 45 pages long saying the same thing with much more information.... So you have 40 pages that concern diagnosis and information (a lawyer practicing and living in one of the study municipalities).

Cases, where the effect of the IAL would be quick to block the process, would be very marginal. The lawyers we met themselves had never had to deal with this situation. According to them, the only cases where the floodability of a property was really so problematic as to block a sale, were the (marginal) cases of old buildings with no building permit.

> I had a file last week, it's been a long time since we've had a case where we've had someone who bought between 1977–1980, who built his house, who sold the land, who sold the house and then came to sell his house here. We took out the risk status and saw that we were in zone R1 of the risk prevention plan, the zone affected by water. We refused to sign the sale without a building permit. They said give us the permit and the certificate of conformity or I won't sell the property. Because it is important to know that as a lawyer, (...) for a house that does not have a permit, that is demolished following a disaster and that it will not be rebuilt because there is no permit, it is the lawyer who pays. It is important to know that this is what we also pay as a guarantor. So we decided that we would no longer sell houses on floodplains, which had been built before, without the permit because there are a lot of them here (a lawyer practicing and living in one of the study municipalities).

However, the systematic transmission of information by lawyers can be questioned by two elements highlighted by the survey. The first was revealed by the investigation of "a negative case." A risk manager from a community of municipalities described the dispute in progress at the time of the investigation between a new resident and a lawyer from one of the municipalities concerned by the study, where the information had apparently not been transmitted very clearly.

> Ah… lawyers! Because in fact lawyers were also reminded that they were obliged to annex the CERFA[14] on the IAL. It wasn't necessarily always done... there were omissions! Precisely here, I have an example [2014]: I have a small mazet[15] that has been identified, and so we had to create a refuge area and even the state services were saying that they would see, but if it was possible they would make an expropriation [...] This house was for sale, it was "made like that," it was just a mazet, which is occupied now [...] Then I saw the announcement, and I was very surprised: a mazet, habitable, something like that... so I alerted the mayor, because there was a change in the renewal of the municipal team in 2014 – so, in February 2014 we alerted the former mayor about this very worrying situation, and here, the house was for sale. And the mayor said to me: "Well, that's fine, send me all the information because I've heard from this family." Since the announcement, the flood risk to this house was obscured, we do not say that it took in water... because in fact it was in the major bed! (risk expert in a community of municipalities of study sites, native inhabitant of the department, about 45 years old).

This case, without sufficing to call into question the fact that the obligation to provide information is respected in the vast majority of cases, makes it possible to put into perspective the nature of the information transmitted, and thus to question its potential effect. Without going so far as to conceal the information, the floodability of the dwelling can be put into perspective by the professionals in charge of transmitting this information.

> Yes, we'll tell him: it's flood-prone, but it's common here... for example owners who bought or were told that it had indeed

14 The Cerfa is a regulated administrative form, an official document, the model of which is set out in an order. Each step corresponds to a Cerfa identifiable by its number. Originally, the term referred to the French public body responsible for preparing and registering these documents, known as the *Centre d'enregistrement et de révision des formulaires administratifs*. Sources: https://www.demarches.interieur.gouv.fr/formulaires.

15 A French regional expression (Provence, and by extension in the South) meaning a farmhouse, country mansion, Provençal style country house. Source: https://www.cnrtl.fr.

> been flooded in 2002... by 50/60 cm while the architect when she came to take measurements she immediately saw that it was not possible, the two neighboring houses were at the same height: 6ft and 5″2! The result: flood houses at 6″2! The person has just bought the house, a little more than a year ago, got into debt, had to create a refuge area, and had no possible bank credit... they were in too much debt already (native resident, about 45 years old, risk expert in a commune).

Professionals, real estate agents, or lawyers, with the exception of one of them, all expressed strong criticism of the mapping of PPRIs. This criticism of official zoning is addressed in light of local knowledge of risks that would attest to their inaccuracy, in a classic opposition between the local actors and the state on this technical risk issue [DOU 11]. The professionals interviewed denounced in particular the negative effects of the PPRI in that it is an obstacle to local development, whereas the level of risk would not justify it.

> ***And you, when you say that it annoys you, that it bothers you, this plan, that it is badly done, concretely, in your profession, when does it bother you?*** For me, it is only to develop the economy. You have land that has never taken on water, that is closed off. I had a client (...) there is the city centre, unable to build. So he did what he could. He wanted to expand in order to develop, to carry out activities. He was refused, because it was on a floodplain. The water would never rise. ***So one of the specificities, if we could see one, of the land market in the areas affected by the risks was that the State is much more intrusive?*** Absolutely. It was the precautions that were annoying, because they were removed and it didn't stop the house from being flooded. Indeed, I fully agree with this idea. It was not really water itself that created obstacles, it was the State, they were not bothered. We know well, they were covered, they took all their precautions, they were comfortable (a lawyer practicing and living in one of the study municipalities).

Local professionals denounced devices whose reality did not correspond to what local people knew in the field. This provision concerning official mapping led them to relativize its veracity to their clients when they transmitted the IAL.

> There is an element of town planning that does not want to go out on a limb for you, therefore does not want to issue certificates, non-local buyers, who are sometimes frightened by it all. So we're here to temper these documents by saying, "Don't worry, we're on a floodplain, but there's never been water here and there never will be." For information, Mount Aigoual[16] is on a floodplain. [...] So we have aberrations like that which come from the administrative side, which is no longer a reality. [...] But as a result, if you cover yourself too much, you have areas that are no longer constructible because of this poor mapping. [...] The agencies do not show it and they are right and know very well that local lawyers, we all have the same opinion on this, we all know the region very well and we know very well that in such a place it will never flood, it is impossible (a lawyer practicing and living in one of the study municipalities).

Thus, the obligation to inform purchasers or tenants, introduced by law in 2003, seems in the vast majority of cases to be respected: information on risks is well communicated to people acquiring or renting a property. Nevertheless, after investigation, we see that the effect of the transmission of this information is related to the context in which it is given. First, and according to the professionals in the sector interviewed, the effect of the IAL can be put into perspective when, according to them, the majority of property transfers take place between people in the territory, and therefore, are supposedly already informed about local risks. This result is consistent with an initial assessment of the effects of the IAL[17] conducted by the General Commissioner for Sustainable Development in 2015, which concluded that "the entry into force of the information requirement reduced the probability that PPRi zoned detached homes would be sold to buyers not residing in the municipality and therefore less likely to be informed of local exposure to natural hazards before the IAL." However, the systematic nature can be qualified by the "negative case" revealed by our survey. Above all, the qualitative elements collected by interview make it possible to qualify more precisely the way in which this information is transmitted... and thus its potential effect. In particular, the case study revealed local challenges to

16 Local mountains' name.

17 Economic diagnosis based on the assumption that the effect of the IAL would be revealed by a price decrease whose results, which are insignificant, call for further studies with a longer return time [MAU 15b].

official risk mapping documents – a point that has previously been largely documented by previous work in various territories [BAY 00], [DEC 97], [DOU 11], [LEB 07]. This tendency for local councilors to be suspicious of the state's risk doctrine may lead these professionals to orally relativize the risk when the document to be signed is transmitted. The transmission of information, although now (almost) systematic, can thus be relativized by the contextual elements, particularly human that qualify its transmission.

In order to question the effect on the people who received this information, we placed ourselves beside recipients by focusing in the following section more broadly on the analysis of the influence of flooding on residential choices.

3.3.1. *The marginal influence of flooding on residential trajectories in exposed municipalities*

The impact of natural hazards, and more broadly of environmental attributes, on residential choices remains largely unexplored from a sociological point of view [AUT 10], [GUE 06]. The few available studies have highlighted in particular the need to redefine the notion of proximity around the register of experience and sensitivity in order to improve the explanatory power of quantitative valuation methods and in particular that of the hedonic price method [GUE 06]. The question of residential choices is further explored in the field of microeconomics, where this methodological framework, associated with the assumption that flooding influences residential choices in that it constitutes a strong constraint on housing, and therefore that IAL has a potentially strong impact on the decision to settle or not [MAU 15b], is dominant.

The analysis of inhabitants' narratives in flood-prone areas in our sample puts this assumption into perspective, which is based on rational logic. If we consider all the inhabitants interviewed (regardless of whether or not their homes were located in flood-prone areas), it appeared that the question of settling, staying, or leaving a municipality exposed to floods was not only a question of the risk equation, or even a minor one in these terms, in light of the other prevailing choice criteria. In general, and as shown more generally by the work on residential choices [AUT 10], it was the criteria relating to lifestyle choices that predominated (location of work, friends, family, solidarity networks, etc.) in the narratives of residents to justify their choice of residence.

What the analysis revealed more specifically about our study area was the prevalence of attachment, regardless of socio-professional categories, age, or gender. The analysis of residential trajectories made it possible to highlight the influence of attachment to place on the choices of the inhabitants[18]. This influence had previously been shown in the neighboring department of Ardèche [CHA 13]. In the Gard department concerned by this study, this issue of attachment was related to the land issue. The analysis of the inhabiting trajectories indeed made it possible to highlight the influence of the transmission of family land ownership in the residential trajectories of the Gard. The choice to reside here, which the inhabitants attributed as much to a relationship with the places (territorial attachment) as to links with other inhabitants (family and/or friends), was often encouraged by the availability of family land, further reinforcing the role played in the final decision by the family dimension, personal history, and links to the territory.

> That's why I'm still in Le Vigan and why I bought my grandparents' house [...] I lived there all my childhood there too. We have memories [...] that are there, so I became attached to this house and I didn't want it to go to strangers (resident about 45 years old, native, left and returned to Le Vigan).

Focuses on the analysis inhabitants' narratives whose homes were directly affected by the flood (i.e. located on a floodplain according to the official mapping of the municipality's PPRI) did not change its content. While some had made the choice to leave following a flood, it was not that which dominated their experiences (direct or reported) of the inhabitants encountered. For the majority of them, this was not a sufficient reason to leave.

> But it's a reason that's only valid in the fall season if you want to every time. When there is the Vidourle river passing by and they say "we have to sell," sometimes this is enough. But we have a beautiful plum tree, we have a neighbor who is sweet... It's stupid to say it, but I think it would break our hearts to leave

18 The analysis of life stories in terms of habitat reveals three main criteria to explain the reasons for their dwelling's location: love of the living environment and/or the surrounding nature, attachment to the place and to those who live there, and choice of a way of life. On the analysis of residential trajectories in flood risk areas, the reader may refer to Chapter 2 of Volume 1 [DUR 18] for more details.

> (resident, about 45 years old, of foreign origin, married to a native, Sommières).

When asked why the risk of flooding – or even its occurrence – was not a sufficient reason to move, residents whose homes were located in flood-prone areas stated (or recalled) the assessment criteria described above (living environment, attachments, and lifestyle) as a positive counterbalance to their choice.

> ***Isn't that a question you asked yourself?*** No. [...] I would have been a little more careful, maybe, I would never have lived there. But I've always liked it. [...] We're still a little far from the city... We're in a quiet place. The houses are pleasant. No. I never wanted to leave (native resident, back for more than 10 years, Le Vigan, evacuated in 2014, flooded the first year of residence in 1982).

However, there is a limit to our sample that must be stated here: the residents who decided to move could not be interviewed. Beyond the testimonies of land professionals attesting that post-flood relocation cases remained marginal, one element of the survey nevertheless makes it possible to put into perspective the numerical importance of such cases that would be missing from our survey. All the inhabitants we met were asked about this point regarding their inter-knowledge network. Out of the total sample (33 inhabitants), only three cases of relocation due to flooding could be reported. Among the eight inhabitants (or couples of inhabitants) expropriated following the 2002 flood in Collias, five said they would have preferred to be able to stay (or rebuild) near the water despite the violence of the 2002 flood.

Thus, the results of our qualitative study by interviews with residents show that the question of risk appeared as secondary in the orientation of residential choices when they faced complexity and all the other criteria influencing choices[19]. Nevertheless, in order to compare this result with the work in microeconomics that dominates the subject, we question the influence of the land market on these choices in the following section.

19 See Chapter 2 of Volume 1 for more details on the influence of flooding on residential trajectories [DUR 18].

3.3.2. *A flood that does not affect prices?*

According to the actors interviewed (residents, shopkeepers, insurers, local politicians, lawyers, and risk experts), in this territory 29% of which is affected by flooding in urban areas[20], the location of a property in a flood zone only has a marginal impact on the determination of its value, if we exclude extreme cases (e.g. houses that have been very heavily flooded recently). According to the lawyers we met, the direct impact of the flood on land prices occurred only in two marginal cases: for cases of properties located in areas "known to take in water" (expression of one lawyer), and this, independently of official zoning (i.e. those of the PPRI) or when the property concerned does not have a building permit.

What was decisive according to the professionals met (all types of land combined) were above all the other elements that determine the property's property value: its exposure (its exposure to sunlight in particular is a dominant criterion in the landlocked village of Le Vigan), its location (practicality in everyday life, calmness, local amenities, etc.), and its neighborhood (well developed or not; social housing nearby, for example).

The impact of floods on the price of land in the Gard region is illustrated by the demonstrative power of the city of Sommières, which regularly makes the headlines for its floods. However, this city remains attractive (historical wealth, economic dynamism, strategic location, and tourism) and continues to develop. According to the rich testimony of a local expert, local managers expected, for example, that the 2002[21] flood would have a significant impact on transactions in subsequent years. A review of the 2003 and 2004 period made it possible to counter this feeling since neither the number of sales nor the average price of sales fell in the years directly following the flood.

> The impact of a PPRI on the real estate market? We thought very soon after the floods of 2002 that we would see a downturn in the real estate market in Sommières and when we looked at the transfers after 2002, we realized in 2003, 2004,

20 Source: http://www.noe.gard.fr/index.php/le-gard-et-le-risque-inondation.

21 Floods of September 8–9, 2002 throughout the Gard department. The result: 22 deaths, 299 municipalities in the department out of 353 affected, and more than €830 million in damage. In Sommières, the impacts were very significant on the right bank of the Vidourle on recently urbanized backfilled land where housing developments and a supermarket were devastated [RUI 08].

> that all the declarations of intent to sell that we received in the town hall... in regards to price... almost no impact. ***That is, prices have increased as at the national level?*** Yes, yes. We didn't have... houses that were in a little higher subdivision where there was 3 meters of water or 2 m 50 without a shelter level, it was sold at the same price as before. ***And now what?*** No, no. There's no connection (Director General of the services of a community of municipalities to which a studied municipality is attached).

The prevalence of other criteria over that of risk is particularly illustrated by the case study of Le Vigan. According to the experts interviewed, the impact of the flood on price was marginal and concerned only a few cases of houses taxed as "unsaleable" because they were very exposed.

> ***Do your clients ask you about the floodability of the properties they visit?*** They're going to ask us the question, but there will be no impact on the purchase. (...) ***Does it make any difference in terms of price, that is, concretely, a house?...*** Not at all. No, here, we're going to have differences depending on the location of the house. (...) if you will, all the houses that are here, on the mountain, we are facing due south. On the other hand, you take a house that will be on the other side (...). In winter, there is no daylight. For the same house, the same construction, you put it here or on the other side, you will sell the other one very easily, this one, you will have difficulties. ***So in fact, what you're telling me is that it's marginal, in fact, the question of risk on the price, the impact on the price is marginal?*** Absolutely! (native resident and real estate agent, Le Vigan).

When asked about the case of a neighborhood clearly identified as flood-prone, the point of view remained the same: other criteria determine the price.

> Do you think there is no impact on the price? No, they're cheap little houses. It's going to be a budget house. On the same house, you take a house elsewhere. We won't have a subdivision, because in fact, we don't have a subdivision of this type, there's only this one, you take about the same construction. You put it elsewhere, it will be worth more elsewhere than in the subdivision. It has nothing to do with

> floodability. It has to do with the sector itself (Real estate agent working and living in one of the study municipalities, native).

While there has been a decline in the average value of local land in recent years in Le Vigan, this is due to the economic decline of the city, not to the risks to which it is exposed. A couple of residents testified to this fact. For personal reasons, they would like to sell their house to Le Vigan, but have not left because the price of the house would not allow them to buy an equivalent (in terms of attributes, etc.) elsewhere in the region. However, this is not due to flooding (their house is not located on a floodplain) but to the city's unfavorable economic context.

The case of Le Vigan highlighted an ambiguous link between the impoverishment of the centers and exposure to flooding. The impoverishment of the city center is visible through a simple observation (rows of closed shops, dilapidated buildings, etc.). A particularity seems to appear in this sector of Le Vigan, where according to the different points of view of local experts collected, a local culture – that of the Cévennes – which describes itself as "landowners," would lead to a relative rarity of properties for sale. Owners prefer to keep buildings for rent rather than sell them to outsiders – especially in a context of low land markets – but without necessarily wanting to invest in them or make too many concessions on rental levels. Several testimonies from local residents and land professionals converged to attest to the great obsolescence of the apartments available for rent without the rents being very low (it should be noted again that most of the rentals are located upstairs). In comparison, first-floor apartments with gardens, even if they are located in flood-risk areas, can be more attractive.

> I'll show them the apartment, if there's an outside, people will take it. I know I'm going to rent quite easily [...] despite the fact that it is on a floodplain. We're really involved the flood subject, right in the middle of it. ***Because does it have other benefits that make it interesting?*** Because you're going to have an exterior that you won't have in the center, people will find the same rent price, they'll think, "We'll take that." It's not going to be an obstacle at all, because the river is far away. There's going to be more of a panorama, the river, than to say "We're going to flood" (Real estate agent working and living in one of the study municipalities, native).

The question of the links between flooding and the impoverishment of city centers is a complex one. The question of the floodability of the centers is an additional factor, but it is not enough to explain the decline in attractiveness of the old city centers, particularly in this region. The work of Cuenot and Brun [CUE 17] shows that this phenomenon of impoverishment of old city centers – which is not specific to the department but common to the entire Languedoc region–is the result of a complex process combining regional economic history (decline of small local shops in particular), development choices (peri-urbanization in particular), and residents' preferences. At a time when mobility is ensured by owning a car, a typical house with a garden attracts more people than a town house. Households – at least the wealthiest (pensioners, managers and professionals, and employees) – prefer to move away from the center by building a detached house or renovating an old farmhouse in the surrounding area. Taking advantage of a better living environment (calm and local amenities), they continue to have access to all the city's amenities by moving away from it without leaving it completely. At the same time, there is a process of "rural renaissance" – already observed for a long time [KAY 72], where middle classes in search of low-density areas and, for some, alternative lifestyles have preferred rural hamlets, farms, isolated mills, or even small peripheral residential developments to the historic centers of small towns [CUE 17]. While floodability alone cannot explain the phenomenon of impoverishment of the city centers, it remains that only the latter is very real. The impoverished centers offer housing at a lower cost that attracts the most economically vulnerable populations[22].

However, other observations allow us to put into perspective the observation of an accumulation of economic and social vulnerabilities and exposure to floods. The exposure of these socially vulnerable populations to flooding should be put into perspective, given that most of the housing in question is located on the upper floors, with buildings in the city centers being mainly occupied by businesses below. This point is particularly visible in the case of the city of Sommières, with a city center located in a zone of uncertainty, but remaining a strong tourist attraction. Almost all the first floors of the center are occupied by shops. Several testimonies corroborate

22 A limitation to our sample must be highlighted here: the most economically poor social groups were also the most difficult to meet. Our analysis, therefore, suffers as a result: the impossibility of including the point of view of these inhabitants in a situation of economic vulnerability. Attempts to complete our sample with such profiles of inhabitants, particularly through communal social centers, have been unsuccessful.

that the local land market of these businesses (whose price is usually defined according to turnover) does not suffer from the flood.

> I bought here, I knew it was flood-prone and all that, I had done my homework, I went to a shopkeeper... because we know it, Sommières, it's... I had checked with the people… especially those who had businesses, erm. And then for them, it seemed normal, you know. The floods, okay, we'll go up, remove everything, clean it all up, mop everything up, and go again. So I said, why wouldn't it work the same for me, huh? Especially since I had the floors, knowing that all businesses have one floor, most of them. That's why it can be foreseen. And all this in Sommières, everything has always been sold and everything has always been redeemed. ***You bought at a below-market price, do you think?*** No. ***No, the water issue... doesn't it lower prices?*** No, no (resident merchant of Sommières, settled for more than 10 years, arrived from another region, married a Gard native).

Moreover, in a very unexpected way, a real estate agent answered the question about the effect of the floods on the land market in the weeks following an event by explaining that it could have a positive effect on properties that were actually located on a floodplain but did not flood during the last event.

> Yes, it has an advantage, we arrive, we have a roof, it was after September, October, when there were the big episodes, we arrived, he looked at the house, "That's a good roof." Yes indeed, it hadn't taken in water! (a lawyer practicing and living in one of the study municipalities).

The result of our qualitative study with the inhabitants and real estate professionals in the Gard, a territory with regular floods but where the land market is stimulated by other criteria (landscape, climate, tourism, etc.) shows that house prices do not suffer from risk. This result thus goes against the theory of hedonic prices, which would argue that housing prices within floodplains, all other things being equal, are necessarily cheaper than in non-floodplains since they have a lower environmental quality [PAL 05] and/or in unsafe areas, i.e. they do not ensure the safety of people [BRO 85]. The results of our qualitative study with real estate professionals and residents of flood-prone areas of the Gard are in line with the previous work by interviews, which had similarly shown that the risk factor was not a decisive factor in residential choices and did not have a significant effect on prices [CAU 03], [DER 06], [HUB 03]; at least, while there may be an influence on choices, it remains less so than that of other factors (environmental amenities, emotional ties, practicality of the places, spatial proximity, etc.).

3.4. Discussion

Our sociological study in the Gard department with more than 50 inhabitants and real estate professionals shows that residential choices are made on the basis of criteria other than risk alone. It thus shows an apparent absence of any significant effect of floods on land prices, in line with the analyses of our previous work in the neighboring department of Hérault [DUR 14] and those of previous studies on other territories [CAU 14], [DER 06], [HUB 03]. From the point of view of residential choices, we see here that, unlike other analyses available in the literature on foreign territories, it is not necessarily captive audiences [NOV 12] who do not have the means to do otherwise, who remain in floodplains. From the point of view of the land market, the economic literature mainly suggests two ways to explain the potential effect on prices: the insurance system specific to natural disasters, which does not price risks at their insurance cost, and perception bias [MAU 15a].

Concerning the first, the insurance issue is indeed the one that the real estate professionals we met identified as a probable source of change. To date, according to them, only the issue of mitigation measures (made mandatory by the latest generation of risk prevention plans) can slow down buyers in some cases. In the majority of these, however, measures are not taken and controls do not exist. That being said, and on the understanding that legislation and standards in risk management evolve with disasters (see Chapter 5 of Volume 1), non-compliance with this obligation could be highlighted during future floods and possibly cause lines to shift.

> As long as there is no repeat of 2002 in Sommières, it will highlight that people have not carried out mitigation measures, perhaps on this scale insurers will ask questions... but without that... As long as it affects a parcel of land, it's temporary. This is not a problem (a lawyer practicing and living in one of the study municipalities).

The sociological perspective allows us to refine the second explanatory avenue. Putting things in terms of "perception bias" maintains the illusion of an ideal decision-making model from which individuals would deviate, implying a form of deviation from what the "normal" course of things should be... according to rational economic logic. None of the inhabitants interviewed questioned their municipality's exposure to the floods. While the risk was recognized, it was not enough to give up the other attributes of the territory that justified wanting to remain there. Faced with the probable but perceived uncertain and rare risk, the "force of everyday life" [DUR 18] prevailed over the other assessment criteria, some of which were continuous[23]. Decision-making in "perfect" conditions would require a complete and "objective" representation of risk in a context of certainty. However, we know that this representation cannot exist: in essence, a risk arises in an uncertain context and is socially constructed[24]. While works on the socio-anthropology of risk, widely recognized for their contributions, have been able to feed this interpretation in terms of different representations of risk in the sense of a defined object [DOU 83], [PER 00], [SLO 87], they have since been widely criticized for the cultural determinism they carry [DUC 87]. A *second method of explaining risks* [GIL 03] focuses more on the process of politicization of risks and the diversity of knowledge about them. Work on *the local risk scenes* has shown that risks are not defined unilaterally by the authorities but are produced at the end of processes of interaction and confrontation, in which "non-experts" play as important a role as experts [DEC 97]; others have also shown how much inhabitants can oppose a political approach to their territory focused on the risk of flooding when the meaning it carries in their eyes goes far beyond this issue alone [DUC 03]. In this sense, we defend here the interest of approaches that are more sensitive to the process and that make it possible to reintegrate the question of risk into a living space that is understood in all its diversity [DUR 14].

3.5. Conclusion

Our intention was to propose a sociological point of view to the study of residential choices in high-risk areas, mainly invested by microeconomics.

23 See Chapter 2 of Volume 1 for details of this analysis of "the force of everyday life" on inhabitants' choices.

24 The literature on the social construction of risks is so dense that it is difficult to be exhaustive in terms of referencing: see, for example, Duclos [DUC 87].

In particular, we proposed to question – through the results of a qualitative study on residential trajectories in flood risk areas – the two "objective" elements likely to influence residential choices: the law – through the principle of information for buyers and tenants (IAL) of risk exposure – and the land market. These criteria have been retained as the usual objects of study to define the impact of floods on the housing sector by economists, with the underlying assumption of rational choice.

Our results, discussed in the previous section, lead us to the conclusion that microeconomic theory, and in particular the assumption of perfect expectations of future damage, does not pass the test of our survey. In support of a qualitative study among more than 50 inhabitants and real estate and risk professionals: flood risk does not seem to significantly influence house prices in flood-prone areas[25]. Thus, we have not found any further evidence of a *homo economicus* in floodplain areas than we think exist elsewhere. The theoretical maintenance of this category in microeconomic thought – which dominates the view of contemporary French social sciences in terms of environmental issues in particular – reflects more the hegemony of neo-liberal economic thought than a heuristic interest. Framing by an alleged economic rationality in residential choices does not hold up to the complexity of human logic. Guemard's exploratory work [GUE 06] had questioned the hedonic price method in its ability to understand residents' preferences in terms of residential choices, particularly around an environmental factor of proximity (green space). Herbert's seminal work [HER 97] and his theory of limited rationality already underlined how little the conceptualization of residential choices through the search for objective attributes is relevant and cannot express all the preferences that feed residential choices. In order to better understand the complexity of the social dimension, our work – joining others [DUR 14], [GRA 05], [GUE 06], [ING 00], [LAT 97], defends the interest of studying situations "in the process of being created" by focusing attention on the practices of actors in situations, and by taking seriously the emotional dimension and particularly the attachments that link them to their living environments and the environmental entities that constitute them.

25 Initially, within the framework of the MobiCLIMeX project, we hoped to be able to complete this survey with a longitudinal study of the change in land prices in the municipalities concerned in flood-prone and non-flood-prone areas, but this could not be done due to the lack of reliable data available over a sufficiently long period.

3.6. Reference

[AUT 10] AUTHIER J-Y., BONVALET C., LEVY J.-P. (eds), *Élire Domicile. La construction sociale des choix résidentiels*, Presses universitaires de Lyon, Lyon, 2010.

[BAY 00] BAYET C., "Comment mettre le risque en cartes? L'évolution de l'articulation entre science et politique dans la cartographie des risques naturels", *Politix*, vol. 13, no. 50, pp. 129–150, 2000.

[BER 97] BERTAUX D., *Le récit de vie*, Armand Colin, Paris, 1997.

[BRO 85] BROOKSHIRE D.S., THAYER M.A., TSCHIRHART J. *et al.*, "A test of the expected utility model: evidence from earthquake risks", *The Journal of Political Economy*, vol. 93, no. 2, pp. 369–389, 1985.

[CAU 14] CAUMONT V., Risque de submersion marine et marchés fonciers et immobiliers sur le littoral du Nord-Pas de Calais. Phase 2: analyse quantitative des marchés fonciers et immobiliers, Cerema Report, 2014.

[CHA 13] CHARDONNEL S., DUVILLARD S., SGARD A. *et al.*, "Devenir propriétaire loin des métropoles : entre contraintes de mobilité et choix de vie", in GERBER P., CARPENTIER S. (eds), *Mobilités et modes de vie. Vers une recomposition de l'habiter*, Presses universitaires de Rennes, Rennes, pp.133–158, 2013.

[CGD 14] CGDD-SOeS, "Le sentiment d'exposition aux risques environnementaux", *Chiffres et Statistiques*, no. 567, publication du MDDE, Paris, 2014.

[CRE 18] CREUTIN J.D., SANSEVERINO V., "Legal Evolution of the Conflict Between Water Mobility and the Mobility of People", *Mobility in the Face of Extreme Hydrometeorological Events 1*, ISTE Press Ltd, London and Elsevier Ltd, Oxford, 2018.

[CUE 17] CUENO C., BRUN A., "Le mal logement dans les centres anciens des petites villes, l'exemple languedocien", *Urbanité*, January 8, 2017.

[DEC 97] DECROP G., *De l'expertise scientifique au risque négocié. Le cas du risque en montagne*, Edition QUAE, Versailles, France, 1997.

[DER 06] DERONZIER P., TERRA S., "Les bénéfices économiques de la protection contre le risque d'inondation", Ministère de l'Écologie et du Développement Durable, D4E, *Série études*, no. 06–E05, 2006.

[DOU 83] DOUGLAS M., WILDVASKY A., *Risk and Culture. An Essay on the Selection of Technologicalo and Environmental Dangers*, University of California Press, Berkeley, 1983.

[DUC 03] DUCHENE F., MOREL JOURNEL C., "Risques et dynamiques territoriales : La rivière, un élément du territoire parmi d'autres", *Les Annales des Ponts et Chaussées*, vol. 105, pp. 54–62, 2003.

[DOU 11] DOUVINET J., DEFOSSEZ S., ANSELLE A. *et al.*, "Les maires face aux plans de prévention du risque inondation (Ppri)", *L'espace géographique*, vol. 40, no. 1, p. 31, 2011.

[DUC 87] DUCLOS D., "La construction sociale du risque: le cas des ouvriers de la chimie face aux dangers industriels", *Revue française de sociologie*, vol. 28, no. 1, pp. 17–42, 1987.

[DUR 14] DURAND S., Vivre avec la possibilité d'une inondation? Ethnographie de l'habiter en milieu exposé et prisé, Sociology PhD thesis, Université Aix-Marseille, 2014.

[DUR 15] DURAND S., "Du marécage à la résidence. Trajectoire d'une zone inondable", *Annales de la recherche urbaine*, no. 110, October 2015.

[DUR 16] DURAND S., RICHARD-FERROUDJI A., "Du principe de prévention aux cultures locales du risque inondation", in RUDOLF F. (ed), *Les Villes à la croisée des stratégies globales et locales des enjeux climatiques*, Presses Universitaires de Laval, Quebec, 2016.

[DUR 18] DURAND S., RUIN S., "Social effects of floods on mobilities: A comparison between short-term and long-term perspectives", in LUTOFF C. and DURAND S. (eds), *Mobility in the Face to Hydrological Extrem Events. Part 1 : The Relevant Scales of Snalyse*, ISTE Press Ltd, London and Elsevier Ltd, Oxford, 2018.

[GIL 03] GILBERT C., "La fabrique des risques", *Cahiers internationaux de sociologie*, vol. 1, no. 114, pp. 55–72, 2003.

[GUE 06] GUEYMARD S., Facteurs environnementaux de proximité et choix résidentiels, *Développement durable et territoires*, Dossier 7, 2006, available at: https://journals.openedition.org/developpementdurable/2716.

[GRA 10] GRAFMEYER Y., "Approches sociologiques des choix résidentiels", in AUTHIER J.Y., BONVALET C., LÉVY J-P. (eds), *Élire domicile. La construction sociale des choix résidentiels*, Presses Universitaires de Lyon, Lyon, 2010.

[GRA 05] GRAMAGLIA C., "Mobilisations exceptionnelles contre une usine de batterie au plomb", in REMY E., D'ALESSANDRO-SCARPARI C., NOVEMBER V. (eds), *Espaces, savoirs et incertitudes*, Ibis Press, Paris, 2005.

[HER 97] HERBERT S., *Models of Bounded Rationality: Empirically Grounded Economic Reason*, vol. 3, MIT Press, Cambridge, MA, 1997.

[HUB 03] HUBERT G., CAPBLANCQ J., BARROCA B., "L'influence des inondations et des documents réglementaires sur le marché foncier en zone inondable", *Annales des Ponts et Chaussées*, no. 105, pp. 32–39, 2003.

[ING 00] INGOLD T., *The Perception of the Environment. Essays on Livelihood, Dwelling and Skill*, Routledge, London, 2000.

[JOU 14] JOUBERT X, Questionner le principe de l'information acquéreur locataire en zone inondable, Séminaire de recherche de l'INSET: Prévention et gestion des risques dans les collectivités locales: Quelles gouvernances et modalités face aux risques d'inondation et de submersion marine?, CNFPT de Montpellier, September 4 and 5, 2014.

[KAY 72] KAYSER B., Un système d'information sur les petites villes françaises, Travaux et documents du C.I.E.U, France, p. 125, 1972.

[KAU 96] KAUFMAN J.C, *L'entretien compréhensif*, Nathan, Paris, 1996.

[LAT 97] LATOUR B., *Nous n'avons jamais été moderne*, La Découverte, Paris, 1997.

[LEB 07] LE BOURHIS J.-P., "Du savoir cartographique au pouvoir bureaucratique. Les cartes des zones inondables dans la politique des risques (1970-2000)", *Genèses*, vol. 68, no. 3, pp. 75–96, 2007.

[LON 01] LONGUEPEE J., ZUINDEAU B., "L'impact du coût des inondations sur les valeurs immobilières: une application de la méthode des prix hédoniques à la Basse Vallée de la Canche", *Cahiers du GRATICE*, no. 21 (special edition on real estate economics), pp. 143–166, 2001.

[MAU 15a] MAUROUX A., "Exposition aux risques naturels et marchés immobiliers", *Revue d'économie financière*, vol. 117, no. 1, pp. 91–103, 2015.

[MAU 15b] MAUROUX A., L'information préventive améliore-t-elle la perception des risques? Impact de l'Information Acquéreur Locataire sur le prix des logements, Commissariat général au développement durable, Work paper, French Association of Environmental and Resource Economists, 2015.

[MON 94] MONTZ B.E., TOBIN G.A., "The flood hazard and dynamics of the urban residential land market", *Journal of the American Water Resources Association*, vol. 30, no. 4, pp. 673–685, 1994.

[NOV 08] NOVEMBER V., PENELAS, M. *et al.*, "L'effet Lully': un territoire à l'épreuve d'une inondation", in GRAMAGLIA C. (ed.), *L'eau: un bien commun à composer. Cosmopolitiques*, no. 17, Editions Apogées, Rennes, 2008.

[NOV 12] NOVEMBER V., "Comment favoriser l'équité territoriale face aux risques?", *Métropolitiques*, May 4, 2012.

[PAL 05] PALMQUIST R.B., "Property value models", in MÄLER K.G. and VINCENT J.R. (eds), *Handbook of Environmental Economics*, vol. 2, Elsevier, Amsterdam, 2005.

[PER 00] PERETTI-WATTEL P., *Sociologie du risque*, Armand Colin, Paris, 2000.

[ROS 74] ROSEN R., "Hedonic price models and implicit markets: product differentiation in pure competition", *Journal of Political Economics*, vol. 82, pp. 34–35, 1974.

[RUI 08] RUIN I., CREUTIN J.D., ANQUETIN S. *et al.*, "Human exposure to flash floods–relation between flood parameters and human vulnerability during a storm of September 2002 in Southern France", *Journal of Hydrology*, vol. 361, no. 1, pp. 199–213, 2008.

[SLO 87] SLOVIC P., "Perceptions of risk", *Science*, no. 236, pp. 280–285, 1987.

[TEN 14] TENEAU G., *Empathie et compassion en entreprise*, ISTE Editions, London, 2014.

[TAL 07] TALLON J.M., VERGNAUD J.C., "Incertitude et information en économie de l'environnement. Choix privés et attitudes individuelles face à l'incertitude", *Revue française d'économie*, vol. XXII, no. 2, pp. 3–56.

[THO 11] THOMAS M.P., PATTARONI L., KAUFMANN V., "Modes de vie, mobilité et organisation quotidienne des familles: quelles relations avec les choix résidentiels?", in CARPENTIER S., GERBER P. (eds), *Les interactions entre mobilités quotidienne et résidentielle à l'épreuve des nouvelles pratiques sociales*, Presses Universitaires de Rennes, Rennes, 2011.

[TVE 73] TVERSKY A., KAHNEMAN D., "Availability: a heuristic for judging frequency and probability", *Cognitive Psychology*, vol. 5, no. 2, pp. 207–232, 1973.

4

When Driving to Work Becomes Dangerous

4.1. Introduction

Although flash floods affect relatively small areas compared to other types of floods, they lead to a significant number of victims. Based on EM-DAT ("International Disaster Database") data, Jonkman [JON 05] showed that the mortality rate associated with flash floods is significantly higher than that of slow floods (3.6% vs. 0.5%). The various studies on the human impacts of flash floods have shown a high prevalence of car-related accidents in Europe (40%–50%) and the USA (more than 60%) [ANT 01], [ASH 08], [DIA 13], [FRE 83], [JON 05], [MAP 09], [SHA 12], [SPI 14]. Based on a detailed analysis of the circumstances of the deaths, we also found a link between the type of circumstance (e.g. outdoors or indoors), the victims' profile, the watersheds response times, and the accident times [RUI 08], [TER 16]. Thus, unlike deaths at home, which mainly affect the elderly during the reaction of large watersheds (>1,000 km^2) reacting in several hours, travel-related deaths mainly affect men in their prime, surprised by the very rapid rise of rivers draining small areas (a few dozen km^2). This observation raises questions about the respective influence of exposure factors related to daily mobility practices and behavioral factors in response to the emergency caused by flood situations.

Because it is both sudden and localized, a flash flood seems to surprise victims in the course of their daily lives and abruptly disrupts programmed

---

Chapter written by Isabelle RUIN, Saif SHABOU, Sonia CHARDONNEL, Céline LUTOFF and Sandrine ANQUETIN.

activities that take place in locations that are often distant from each other and generate *de facto* recurrent and sometimes almost mandatory mobilities. This singularity of flash floods calls for the adoption of methods for assessing personal exposures that take precise account of the situation in which people find themselves at the time of the event. However, we do not have adequate tools to estimate social exposure to these hydrometeorological events and more particularly the exposure of road users, who represent a large proportion of the victims. Indeed, the only information we have comes from the collection of impact data or the static assessment of exposure by superimposing flood-prone areas and the location of residential or professional buildings. Thus, in 2013, the Gard flood risk observatory estimated that 35% of the department's inhabitants live in flood-prone areas[1] and 14.5% of the Gard companies[2] are located there. Nevertheless, these statistics hide significant territorial disparities and do not take into account the fluctuation of human exposure during the day due to daily travels. Thus, it is currently very difficult to estimate the number and characteristics of people exposed to these risks without taking into account the interaction between the physical and social dynamics of exposure.

At hydrometeorological level, several advances have been made in the development of tools for observing and modeling the spatio-temporal dynamics of flash floods in the Cévennes region and the Gard department. Distributed hydrological modeling tools have been developed to simulate hydrological responses over the very small basins characterizing these regions [ANQ 10], [BRA 10], [VAN 16]. In addition, computer-based simulations in accordance with the identification of road closures during past events have made it possible to better predict the flooding of road sections during such events [NAU 13], [VER 10]. From the point of view of modeling mobility behavior, the needs in terms of planning and managing transport demand have led to the development of the so-called "activity-centric" models to simulate daily travel at the individual level. It is on the basis of these methodological advances that we have created our approach to modeling motorists' exposure to flash floods in the Gard, one of the French departments most frequently affected by these dramatic episodes. Indeed, Since 1225, the Gard department recorded 511 floods and 65% of the 353 municipalities have suffered at least 10 of the referenced floods, some of which have even been affected more than 100 times[3].

1 http://www.noe.gard.fr/index.php/observatoire-du-risque-inondation/indicateurs?id=43.

2 http://www.noe.gard.fr/index.php/observatoire-du-risque-inondation/indicateurs?id=46.

3 http://www.noe.gard.fr/index.php/observatoire-du-risque-inondation/indicateurs?id=56.

At the interface between hydrometeorological extremes and daily mobility, this chapter analyzes the physical and social spatio-temporal dynamics that are at the origin of human exposure to rapid flooding in the Gard. We first look at characterizing each of these dynamics separately before considering their interactions. The study of these interactions is made possible by the development and implementation of the *MobRISK* model (see section 4.4 of this chapter), which is a simulation tool for simulating motorists' exposure to road flooding developed in the Gard department. The analysis of the results of the *MobRISK* simulations finally makes it possible to quantify the exposure of motorists to road flooding during the floods of September 8–9, 2002 in Alès to identify the profile of the individuals most exposed to this event and to test the effect of perception and risk assessment variables on the exposure level.

4.2. Spatial and temporal dynamics of road network exposure to flash floods

In characterizing motorists' exposure to road flooding, the first step is to assess the exposure of the road network to flash flooding. By exposure, we seek to locate the submersible points of the Gard road network and to understand under which hydrometeorological conditions they are likely to be flooded.

4.2.1. *Sensitivity of the Gard road network to high water levels*

The Gard has a 12,321-km-long road network composed of four types of roads: local (84%), main (5%), regional (10%), and highways (1%). In regards to this road network, we must consider the hydrological network composed of a little more than 7,087 km of watersheds (according to the IGN's BD–CARTHAGE[4]), nearly 90% of which has drainage surfaces less than 100 km^2, i.e. likely to cause flash flooding [DEB 16]. It is at the intersection of these two networks that the sections of floodable roads are located, which, depending on the intensity and spatio-temporal dynamics of the rain event under consideration, will be more or less likely to be flooded.

Versini *et al.* [VER 10] and Naulin *et al.* [NAU 13] have developed a method to locate and classify the susceptibility of road/river intersections to

4 *Institut National de l'Information Géographique et Forestière* (French National Institute for Geographical and Forest Information) – www.ign.fr.

flooding throughout the department. To do this, they relied on the PICH (Plan d'Intervention Crises Hydrologiques) database, which lists 40 years of observations of road closures due to flooding carried out by network maintenance agents in part of the department. They identified 1970 potential cut-off points divided into four classes of very low (S_0) to high (S_3) susceptibility composed of 1093 (S_0), 359 (S_1), 297 (S_2), and 221 (S_3) points, respectively (Figure 4.1). The lowest susceptibility refers to road/river intersections that are likely to be flooded for an empirical return time of more than 40 years. The "low" (S_1), "medium" (S_2), and "high" (S_3) sensitivity levels correspond to floods with empirical return times of less than one year for 20%, 35%, and 65% of the points, respectively [DEB 16].

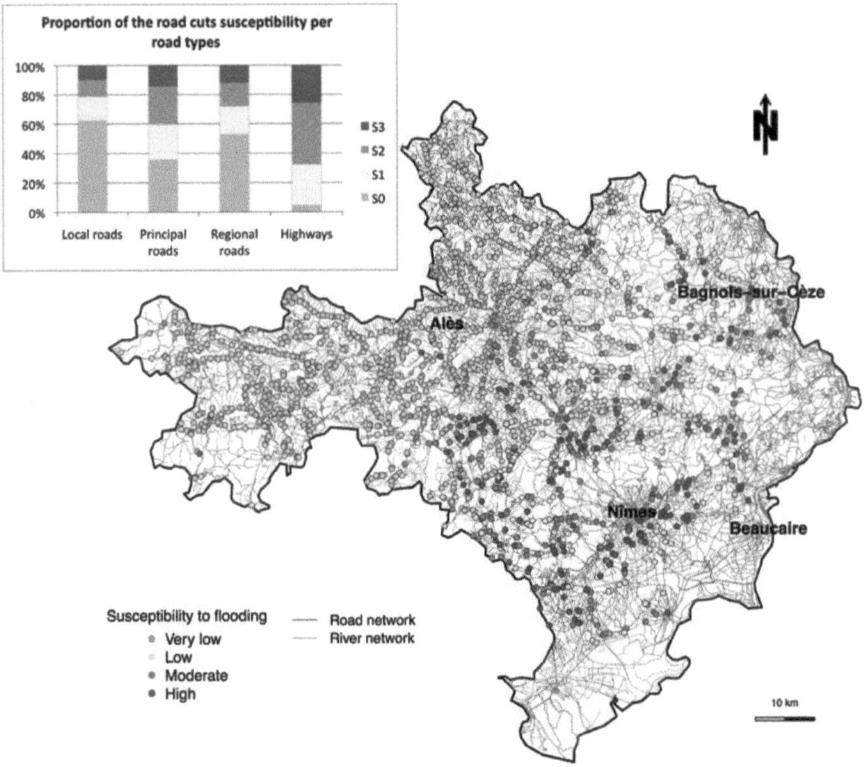

Figure 4.1. *Spatial distribution of potential cut-off points according to their susceptibility level. The diagram shows the distribution of susceptibility levels by road type. Source: [SHA 17]*

The spatial distribution of the potential cut-off points shows that the average to high flood susceptibility points are mainly located in a window bounded to the north by the cities of Alès and Bagnols/Cèze and extending south to beyond Nîmes (Figure 4.1). The majority of cut-off points (60%) are located on local and regional routes (27%), which seems logical since these two types of roads have the highest linear mileage. With regard to the number of potential cut-off points per kilometer of road, regional roads (0.45 pt/km), main roads (0.39 pt/km), highways (0.31 pt/km), and local connections (0.11 pt/km) are the most exposed, in decreasing order. Focusing on the highest susceptibility levels (S_2 and S_3), highways and main roads have the highest exposure in proportion and cut-off points/km.

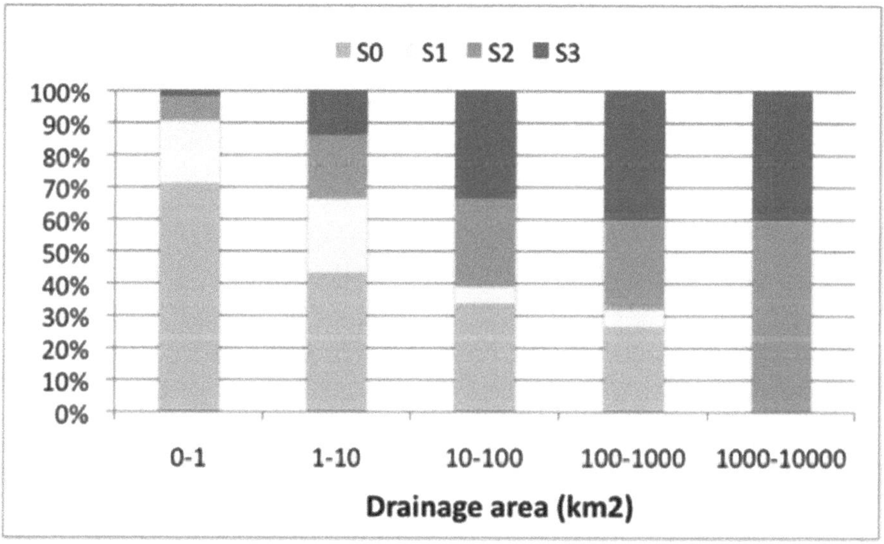

Figure 4.2. *Distribution of potential cut-off points by drainage area and susceptibility level For a color version of this figure, see www.iste.co.uk/lutoff/hydrometeo2.zip*

Another way to describe road exposure is to look at the sizes of watersheds associated with potential cut-off points in order to assess their sensitivity to severe flooding (with few visible warning signs) through the very rapid response of a small watershed. Concerning the distribution of cut-off points by drained area, we observe that the majority of cut-off points are drained by the smallest watersheds: 83% belong to watersheds of less than 10 km^2, with a theoretical response time of less than 15 min [CRE 13]. Of these, 16% have a medium to high susceptibility to flooding (S_2 and S_3).

Medium-sized basins (10–100 km^2) also have a high proportion of medium to high susceptibility cut-off points (67% representing 51 points in total). As for the large (>1,000 km^2), there are only five potential cut-off points, all of which are highly sensitive (S_2 and S_3) (Figure 4.2).

With respect to this static view of the sensitivity of the Gard road network to flash flooding, it is also necessary to consider the flooding dynamics of potential cut-off points on the basis of past rainfall events.

4.2.2. *Precipitation dynamics and flooding probabilities*

We are interested here in the fluctuation of exposure of the Gard's road network by studying three rain scenarios corresponding to real episodes for which (i) spatial and temporal distributions and rain intensities are quite different; (ii) we have submergence cut-off data observed by the road management services. These episodes are those from September 8–9, 2002, September 6–9, 2005, and October 20–23, 2008, which according to observational data caused 136, 86, and 43 roads, respectively, to be submerged in several peaks, on different weekdays or weekends and times, likely to have a greater or lesser impact on traffic (Figure 4.3).

For these three events, flows at the potential cut-off points were simulated using the distributed hydrological model CVN [ANQ 10], [BRA 10], [VAN 16] to obtain a flood chronicle at different points. The simulated flows are then transformed into submergence probability chronicles by taking into account the level of susceptibility of each of the points considered (Figure 4.4) [NAU 13]. These probability chronicles are input data for the MobRISK model.

After characterizing the dynamics of road network flood exposure as a function of hydrometeorological conditions, we will now focus on describing how individuals use the road network on a daily basis.

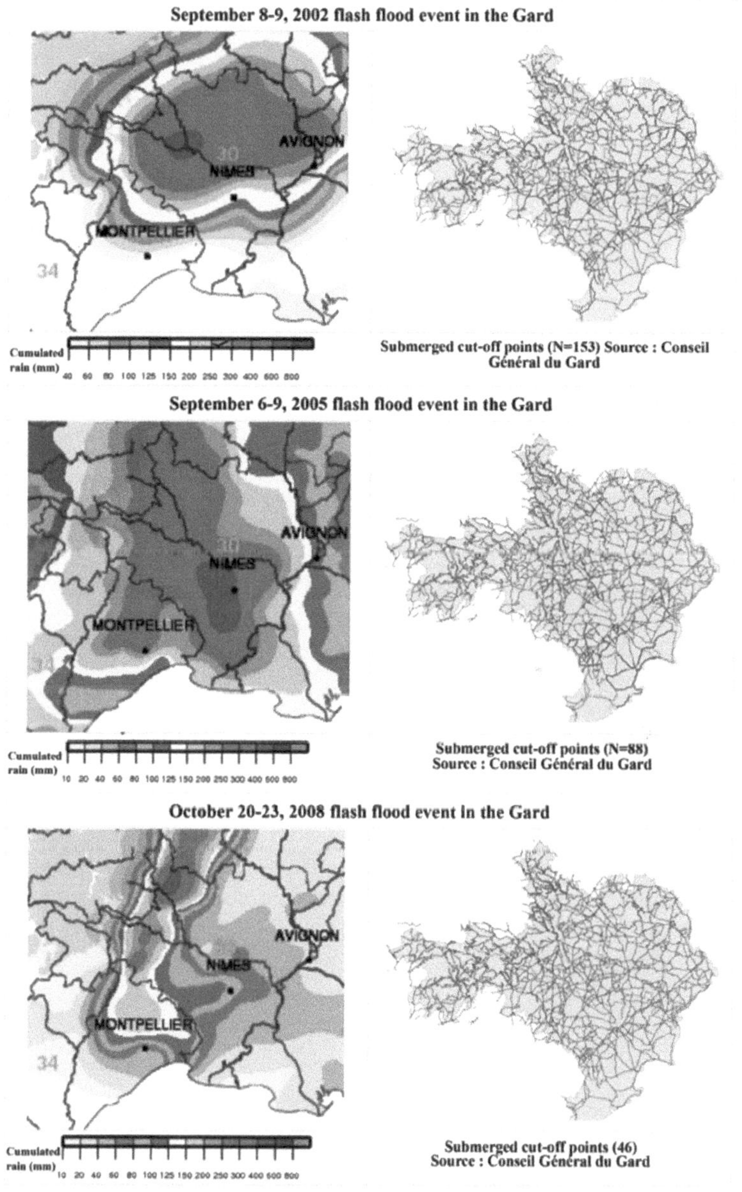

Figure 4.3. *Distribution of cumulative rainfall over the episode (left) and road sections observed as cut (right, green) by the Gard Road Management Service during the three events studied as part of the MobiCLIMeX project. Source: http://pluiesextremes. meteo. fr and CD30. For a color version of this figure, see www.iste.co.uk/lutoff/hydrometeo2.zip*

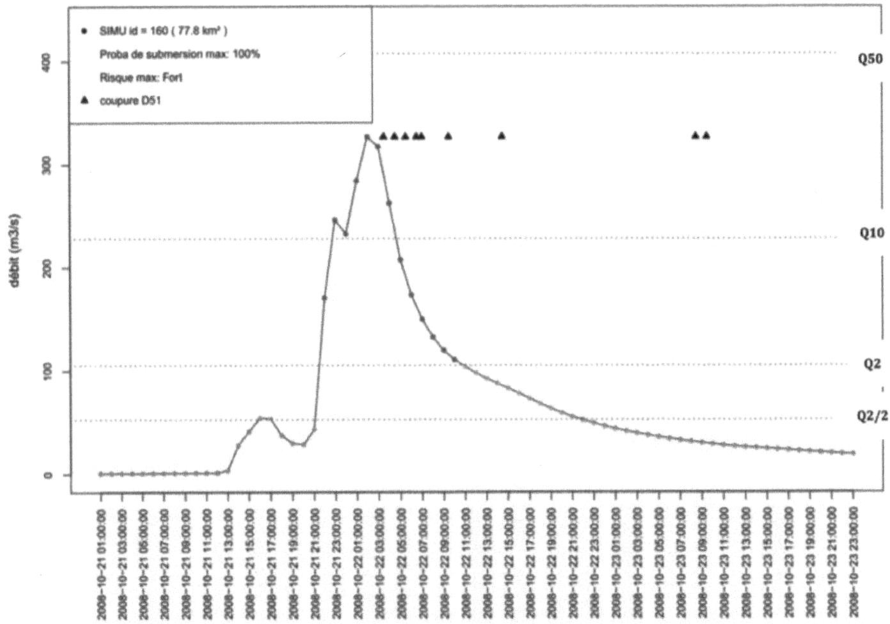

Susceptibility	Simulated peak flow				
	$Q < Q_2/2$	$Q_2/2 < Q < Q_2$	$Q_2 < Q < Q_{10}$	$Q_{10} < Q < Q_{50}$	$Q > Q_{50}$
High	0%	0 à 67%	67 à 100%	100%	100%
Medium	0%	0 à 33%	33 à 57%	57 à 61%	61 à 100%
Low	0%	0 à 20%	20 à 34%	34 à 35%	35 à 100%
Very low	0%	0%	0%	0%	0 à 100%

Figure 4.4. *Example of a simulated flow history for a potential high susceptibility cut-off point. The color points correspond to the probability classes assigned to each time step (discrete). The table gives the correspondence between the return time of the flows (annual to 50 years) and the probabilities of submergence according to the level of susceptibility of the point considered. For a color version of this figure, see www.iste.co.uk/lutoff/hydrometeo2.zip*

4.3. Spatio-temporal dynamics of daily mobility

The objective of exploring the dynamics of mobility is to provide the necessary knowledge to model the "normal" daily mobility conditions of the Gard population, before simulating the exposure of populations to extreme hydrometeorological conditions. The daily mobility of population results from the needs and desires to access activities and resources that are sparsely distributed across space for limited durations and times. Daily travel links activity locations and is part of the general planning of individuals' schedules. Thus, understanding where, when, and how the population moves requires an analysis of the regularities and variations in individuals' activity programs. The numerous studies [HAN 81] that analyze daily mobility behaviors identify three main categories of factors that influence the choices individuals make to organize their daily travel:

– Factors related to the individual's socio-demographic characteristics (age, gender, occupation, etc.) and the household (number of children, family status, number of cars per household) [AND 16], [ETT 09], [LU 99].

– Factors related to the environmental context, such as the spatial (network configuration and traffic condition) and temporal (opening and closing times of services/stores/jobs) accessibility of the activity sites [FOS 16].

– Factors related to the flexibility of activities: [DAM 81] defines the flexibility of an activity as the degree to which the implementation of the activity can be modified spatially (change of place of activity) and temporally (change of time and duration of the activity). As a result, some activities are given higher priority than others and are then planned in the first place and can hardly be modified or canceled.

Several data sources are used to characterize mobility behaviors in our field of study.

First, we mobilized a national survey (*Enquête Nationale Transports et Déplacements* – ENTD) carried out in 2007–2008[5], which includes a regional extension Languedoc-Roussillon, i.e. a sub-sample representative of the population of this region[6]. This first source allowed us to define standard daily mobility profiles because the ENTD describes the socio-demographic characteristics of individuals and their households as well as all trips made, within an 80 km radius of the place of residence, on a weekday and a weekend during the week preceding the survey day.

Then, we used data collected at the municipal level in a comprehensive way to describe the home-work and home-study shuttle transport (MOBPRO and MOBSCO, INSEE files) for all the municipalities in the Gard. This second source was used to spatialize more precisely the geography of

5 SOeS, INSEE, INRETS, national transport and travel survey 2008.

6 The ENTD is based on a representative sample of the entire population of metropolitan France (17,500 dwellings, i.e. a survey rate of 0.056% nationally representative, based on 1999 census data). Languedoc-Roussillon, thus, benefits from a sample of 1,240 individuals.

movements (flows) at the scale of the department on which the case study developed in section 4.2 was located.

4.3.1. *Daily mobility behaviors in Languedoc-Roussillon*

In Languedoc-Roussillon, individuals make an average of 3.7 trips for a weekday and 3.2 trips on a Saturday or Sunday. Statistical analysis of the data from the Languedoc-Roussillon sub-sample of ENTD allows us to distinguish which socio-demographic situations seemed to favor the increase in the individual mobility rate. While the mobility rate for women was higher than that for men during the week, the trend reversed at weekends. The 29–45 age group, on the other hand, was largely the most mobile, both during the week (4.35) and at weekends (3.49). Occupational status must also be added to the factors influencing individual mobility, with working and unemployed people exceeding four trips on average during the week, as work-related activity and job searches being potential reasons for increased travel. Finally, the size of households should be seriously considered to explain mobility rates as the largest households (more than two people) were ahead of the others, weekday and weekend. Again, the reason was the increased need to travel when managing children's activities, for example.

Concerning the reasons for travel, 40% of trips were related to returning home, about 20% of daily trips were for work or study, and the rest were divided between reasons related to purchases, care, administrative procedures, and leisure or visits. This distribution varied, not surprisingly, between weekdays and weekends: travel related to work, studies, care, support, and administrative procedures was more frequent on weekdays; on the contrary, travel related to shopping, leisure, and visits was more frequent on weekends.

Regarding the duration of activities, we noted that individuals spent an average of 18 hours a day at home during the week and 19 hours a day at weekends. Transport times seemed to be similar between a weekday and a weekend. However, they varied according to individuals' socio-demographic characteristics. During the week, men, young adults, and students spent more time on transport. The overall difference between weekday and weekend travel times was mainly explained by occupation (Figure 4.6). By looking at all the sequences of movements and activities (deduced from the reason between two movements) of the individuals surveyed, we can account for the temporal distribution of activities during the day. Thus, Figure 4.5 shows the temporality of activities outside the home carried out mainly before and after lunchtime break. Travel partially followed this pace, since there is a peak in the early morning and late afternoon during the week, but not as much at lunchtime break (effect of the continuous day and catering at the workplace). Weekend travel, on the other hand, is more spread over the whole day and shifted to later hours, both in the morning and at the end of the day.

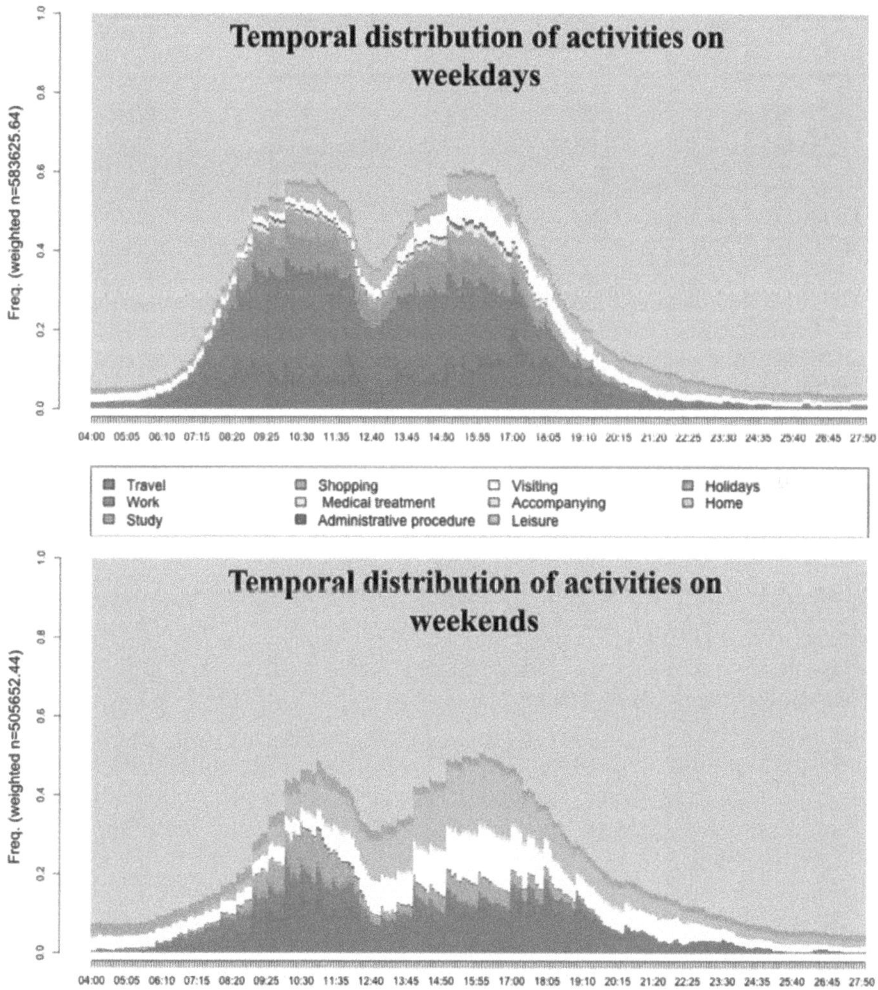

Figure 4.5. *Temporal distribution of activities on weekdays and weekends in Languedoc Roussillon (ENTD data 2007–08, Source: [SHA 16]). For a color version of this figure, see www.iste.co.uk/lutoff/hydrometeo2.zip*

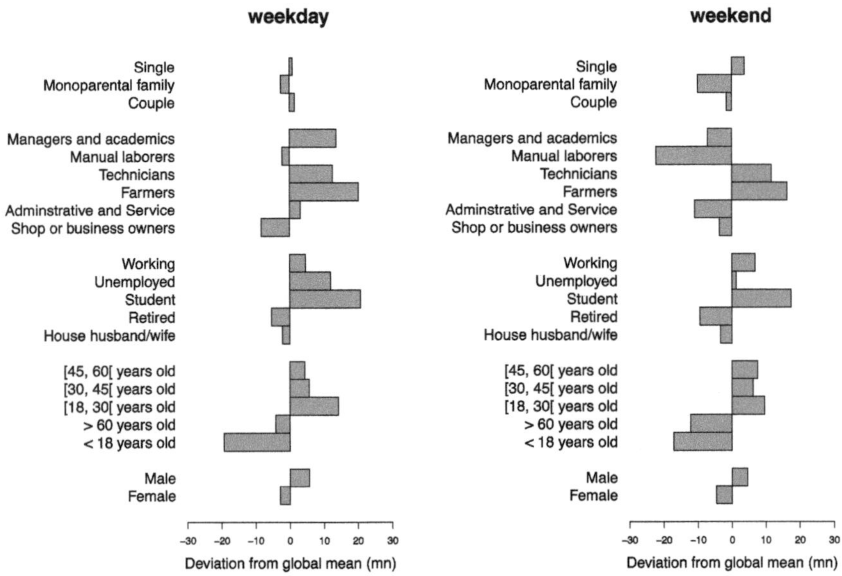

Figure 4.6. *Difference in relation to the average travel times by socio-demographic category for a weekday (D = 18 mn) and a weekend (D = 20 mn). Source: [SHA 16]*

In order to better understand how different categories of people organize their schedules, we conducted a final analysis to compare the sequence of activities and travels of individuals in the 24 hour survey (activity sequences). Thus, we sought to identify types of timetables based on an analysis of sequence dissimilarities based on eight socio-demographic variables. This analysis was followed by a classification (by regression tree) of these sequences [SHA 16].

A first typology concerned weekday schedules. Eight categories emerged from the analysis, which divided two major forms of organization: one centered on the main activity such as work or studies and the other without this structuring activity.

In the first form, students/schoolchildren can be clearly distinguished. Working people differed according to their age category, with younger people working longer hours and traveling earlier. Among young working people, the most educated made longer trips. Students and schoolchildren had similar schedules for the "study" activity, but students had the particularity of having more varied activities and longer trips, especially in the evening (leisure).

The second form of time included "non-active" people. While non-active men traveled longer distances and spent time enjoying leisure activities, women (including housewives and unemployed women were over-represented) spent time "visiting" and "buying" things. Finally, retirees maintained visiting activities, more than leisure activities and tended to return home earlier at the end of the day.

Six types of schedules were available on weekend days. While the youngest people (18–28 years old) devoted their weekends to leisure, the others were scattered among weekend workers (farmers, managers, laborers shopkeepers, and academics) and those who did various activities but differed according to their professional status.

4.3.2. *The daily mobility pattern of workers in Gard*

In order to complete the knowledge of the conditions of daily mobility in our study area, we analyzed the daily mobility pattern of Gard working population who had to reach their workplace everyday.

At the departmental level, nearly two out of three working people moved outside their commune of residence to work. The vast majority of them were heading toward four employment centers: Nîmes, Alès, Bagnol-sur-Cèze, and Uzès. These commuters were mainly men, aged between 25 and 49 years old and were most often employed.

Among employees working in the urban area of Nîmes, 56% worked outside their commune of residence, compared to 66% in the urban area of Alès. Compared to the urban area of Nîmes, commuters living and working in the urban area of Alès traveled relatively short distances to work. Half of these commuters had less than 7 km to travel and a travel time of 15 minutes compared to 11 km and 25 minutes in the urban area of Nîmes, which is more spread out. The categories of employees and women were over-represented among commuters in the Nîmes urban area, while in the Alès area, the proportion of workers was higher. The rural component was not negligible in the Gard, as 3 out of 10 commuters work in these rural areas compared to an average of 13% in France as a whole.

The daily mobility described above allowed us to consider normal travel conditions. From this first analysis, we could see that the number of people traveling fluctuated according to the time of day, but also according to the day concerned (week or weekend). The exposure on the road could not, therefore, be summarized as a single average value. In addition, our analyses show that the number of trips and, even more so, their duration were correlated with people's socio-demographic profiles. It is therefore important to consider this dual-factor (number and duration) when estimating which categories of people are most exposed on the road.

Based on these initial conclusions on normal mobility conditions, we must now consider how during flash floods people are disrupted by the occurrence of heavy rain and the potential flooding of some sections of the road network. In the following section, we look at the factors that influence the adaptation of individual behaviors to environmental conditions.

4.3.3. *Adaptation of mobility behavior in the face of risk*

Several factors are involved in decisions to adopt protective behaviors in the face of extreme events and to adapt mobility behaviors. These factors can be grouped into two categories: contextual factors (which represent observable elements and information received by individuals) and cognitive factors (which are related to the internal processes of processing such information at the individual level) [TER 15].

Contextual factors refer to the different environmental cues and social indices perceived by individuals at the time of the crisis [LIN 12], [RUI 14]. Thus, the reactions of individuals to flood events depend, for example, on the intensity of precipitation and the level of flooding of roads. The analysis of people's narratives from feedback survey data for the Var 2010 event identified three types of external signals perceived by individuals: environmental (precipitation, runoff...), social (individual and collective reactions observable during the crisis), and material (related to flood damage such as washed away cars). The cross-referencing of the types of signals perceived and the change in levels of concern reported by the respondents show that the perception of social signals and material damage considerably increases the anxiety felt by witnesses to the flooding. As part of the mobility behavior, visibility on the roads (which refers to the timing of the event: day or night) also has an important effect on travel choices [KIL 07]. Contextual factors also include warning information received by individuals that may come from different sources (media, social networks, etc.). Terti *et al.* [TER 15] underlined the importance of the context of people's mobility at the time of the crisis in terms of places and activities (what they called "coupled place-activity").

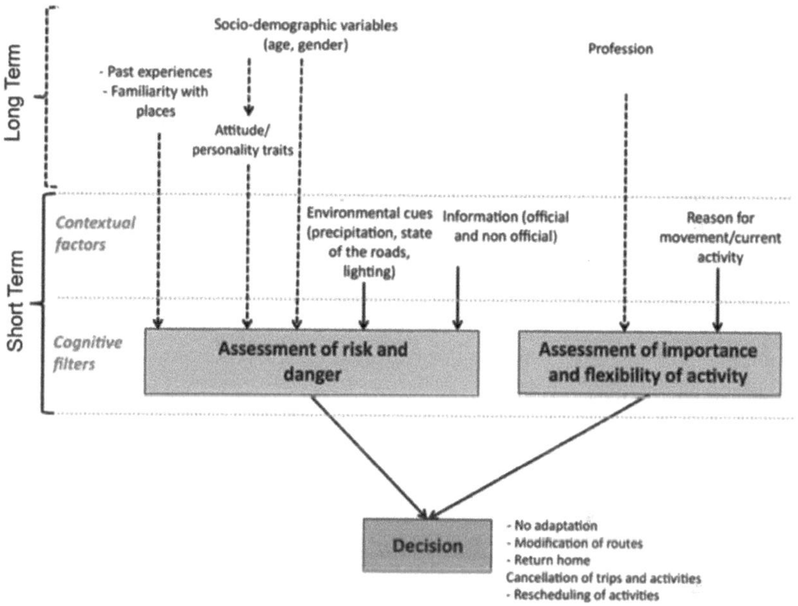

Figure 4.7. *Conceptual diagram of the factors and processes involved in mobility decisions in response to the risk of flash floods. Sources: [SHA 16]. For a color version of this figure, see www.iste.co.uk/lutoff/hydrometeo2.zip*

Cognitive factors are also involved in behaviors since the assessment of the danger associated with perceived environmental signals varies from one individual to another. Travel decisions during flood events can be considered according to two processes: the assessment of the threat and danger associated with the action of moving and the assessment of the importance and flexibility of the reason for mobility (Figure 4.7). Several studies have shown that risk assessment and risk perception depend on people's socio-demographic characteristics. Ruin [RUI 07] has found through surveys of risk perception and reported behaviors associated with vehicle use in high-flood situations in the Gard that men and young motorists tend to underestimate the danger and therefore adapt their mobility less successfully. Other socio-economic variables seem to play an important role in the perception of flood risks such as education level and family and professional status [BOT 09], [LIN 08]. Concerning the evaluation of flexibility, the importance of the reasons for travel and the current activity, the various studies mainly show that individuals have more ease in canceling and modifying the so-called secondary activities (leisure, purchase activities) than work and study activities [COO 13], [RUI 14]. Ruin [RUI 07] points out that the ability to cancel trips depends on professional status, with workers more likely than others to maintain their trips despite amber and red warnings to rain-induced flooding.

In the previous sections, we have examined separately the physical and social dynamics that condition the exposure of daily mobility to high water levels, focusing more particularly on the Gard department in the Occitanie region. However, at this stage, we lack a tool to cross these dynamics in order to assess the number and identify the profile of people likely to encounter flooded roads for a given rainfall event. It is this simulation tool that we will now describe, as well as the case study and the results to which it has led.

4.4. Simulation of the dynamics of human exposure to flash floods

In order to predict and prevent the human impacts of flash floods, it is important to consider individual exposure no longer in a static way, through the location of places of residence, but dynamically, taking into account both water and human mobility at a time step compatible with the rate of occurrence of flash floods. Thus, by developing a model simulating the conjunction of these dynamics, we hope to answer the following questions:

– How many road users are exposed to the potential flooding of sections of road on their daily journeys?

– How does this exposure vary according to the spatial and temporal distribution of rainfall?

– What are the socio-demographic profiles of the most exposed motorists?

– What are the effects of different adaptation behaviors (returning home vs. stopping in front of a submerged road or delaying the time of travel) on motorists' exposure levels?

– What are the effects of different types of attitudes (low- or high-risk aversion) and risk assessment on exposure?

4.4.1. *Simulation method*

MobRISK is a tool for simulating motorists' exposure to road flooding, developed on the scale of a French department (currently Gard). This tool combines three main components (Figure 4.8):

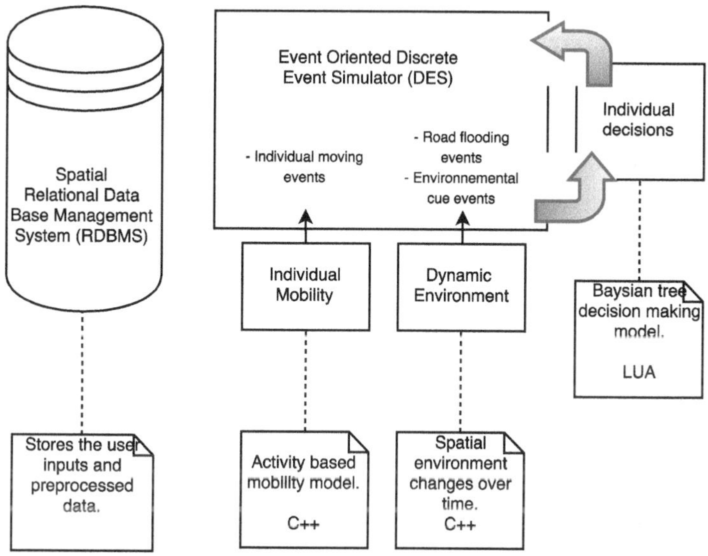

Figure 4.8. *Structure of the MobRISK model. Sources: [SHA 16], [SHA 17]*

– an activity-based mobility module that generates individual road trips based on daily schedules assigned to individuals residing in the study area;

– a module to simulate environmental changes such as precipitation level, probability of road flooding, alert level, etc.;

– a decision-making module that generates individual behaviors for adapting daily mobility to perceived environmental changes;

– a relational spatial database to store the input and output data of the model.

4.4.1.1. *The different modules*

The *mobility simulation module* makes it possible to assign, according to the socio-demographic characteristics of each individual, a timetable composed of a sequence of activities (types of activities and associated schedules) for a weekday or weekend day, similar to the simulated rain event. Then each individual is linked to the different locations where they are likely to perform their time schedule activities. The location of places of residence is relatively accurate, thanks to the socio-demographic and spatial data sets provided by INSEE's population censuses and localized tax revenues. For activities, the data from the complementary exploitation of the INSEE population census enable to only locate the "work" and "study" destinations at the municipality level. The hypothesis of proximity with the main town of these municipalities is made to obtain a more precise location. For other activities (leisure, shopping, etc.), no data is available and we choose to locate these activities randomly in a defined area around the municipality of residence. To move people from one place of activity to another, the route calculation, using the shortest path in time, uses the Dijkstra algorithm [DIJ 59].

The *decision module* enables testing of different behavioral scenarios on the exposure of individuals to road flooding. It is triggered when an individual encounters a submerged stretch of road, receives warning information, perceives a deterioration in environmental circumstances, or terminates an activity before moving on to the next activity. Modeling is based on Bayesian networks to relate the variables involved in decision-making (network nodes) to the causalities between the selected variables (Bayesian network graph structure). The quantification of these causal relationships is defined by conditional probability tables (Figure 4.9). The approach adopted considers that decisions to adapt mobility behavior correspond to three main components: (i) risk assessment, (ii) assessment of flexibility and priority level of the next planned activity, and (iii) decision rules adopted [COO 13], [RUI 07], [TER 15]. Four decision rules are implemented in the model: continue without adaptation ("KEEP CURRENT"), stop in front of a cut-off point ("STOP"), delay the next activity ("DELAY"), return home ("HOME").

The *environmental change management module* consists of a discrete event simulator (DES) that represents the state changes of a given system as a sequence of discrete events. That is, instead of representing the state of the system in continuous time, the DES only takes into account the moments when a change in the state of the system occurs. Six types of changes of state or events are considered: submergence of roads, movement of individuals, crossing of cut-off points, end of activity (which marks the initiation of a movement to the location of the next activity in the timetable), perception of environmental signs (e.g. change in precipitation level), weather warning or vigilance information.

To operate the simulator uses, two types of input data stored in a relational spatial database.

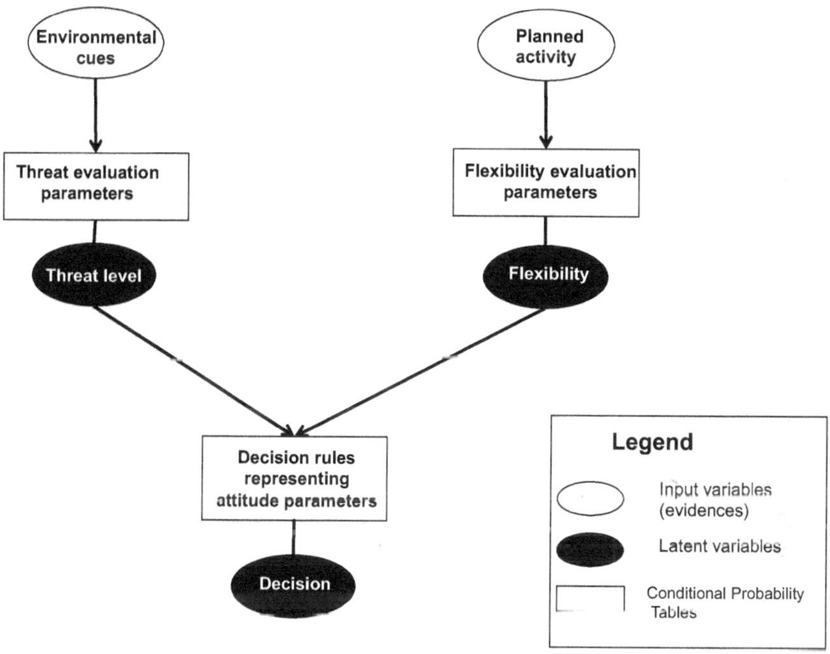

Figure 4.9. *Example of the structure of the Bayesian network used to model the individual decision to start a car journey based on risk assessment (itself conditioned by the perception of environmental signs: precipitation level and light conditions) and the assessment of the flexibility level of the next planned activity. Sources: [SHA 16]*

4.4.1.2. *The relational spatial database*

On the one hand, it is composed of social data provided by INSEE:

– the socio-demographic characteristics of individuals and households residing in the field of study (census data of the population);

– home-to-work and home-to-study mobility data for working people and students (MOBPRO, MOBSCO);

– the daily schedules (sequences of activities on weekdays and weekends) of a sample of individuals who responded to the National Transport and Travel Survey (ENTD).

Other spatially referenced data are needed to combine individual data, travel routes, and hydrometeorological conditions:

– administrative entities (GEOFLA database of the IGN);

– the places of residence of households deducted from the number of households per 200 m^2 (INSEE Localized Tax Incomes-RFL data);

– road and hydrographic networks (BD Carto and CARTHAGE of the IGN);

– the spatial and temporal distribution of rainfall combining radar and rainfall data (re-analyses OHM-CV at 1 km^2 resolution at 15-min time steps) are input data from the behavior model used to assess the visibility and, therefore, risk assessment;

– potential cut-off points (road/river intersections) and their level of sensitivity to flooding (IFSTTAR data from the PrediFLOOD[7] project) [NAU 13], [VER 10].

– simulated flows at potential cut-off points, derived from the simulation outputs of a distributed hydrological model. In this study, we use the outputs of the CVN model using the *LIQUID* modeling platform [ANQ 10], [BRA 10], [VAN 16].

The *input database* developed as part of the MobiClimEx project makes it possible to simulate the events that affected the Gard on September 8–9, 2002, September 6–9, 2005, and October 20–23, 2008, each representing a very different spatial and temporal distribution and rainfall intensity. The addition of other real or virtual events affecting the Gard is easily possible when a new hydrometeorological dataset (rainfall and flows) is provided. On the other hand, if the database structure remains the same, implementing the model in another French department requires the production of another complete data set.

Scenario construction consists of determining experimental schemes by varying two types of model input parameters: (i) the choice of

7 Previous ANR MobiCLIMeX project for distributed flood forecasting for road management in the Cévennes-Vivarais region.

hydrometeorological event to be studied (spatial-temporal distribution of rainfall flows) and (ii) decision scenarios by changing the conditional probability tables of the risk assessment, activity flexibility, and standard attitude variables.

As part of the first implementation of the model described in the following section, we simulated the exposure of motorists in the Alès region during the Gard floods of September 8–9, 2002.

4.4.2. *Results on the case study of the flooding of September 8–9, 2002 in the Alès (Gard) sector*

The case study concerns an area of 39 rural municipalities and 22 urban municipalities located around Alès where 111,511 people live. The sector includes 738 potential cut-off points with very low to high sensitivity to flooding, the distribution of which is representative of that of the entire Gard department. During the 2002 floods, many roads were submerged and one motorist was killed in this area. The objective of the MobRISK simulation on this case study is to quantify the exposure of motorists during this event with regard to the single death and to identify the factors that could have increased or decreased this exposure (timing, perception, and behavior).

4.4.2.1. *Exposure of daily mobility to road flooding*

First, we evaluated individual exposure without taking into account the adaptation behaviors, considering that all individuals maintained their activity program and movements without modification for the two days of the event. The MobRISK simulation launched for these two days, including a weekday and a weekend, generated 737,135 trips with an average number of trips per person of 3.06 on Sunday and 3.64 on Monday.

Figure 4.10 shows the temporal dynamics of traffic at potential cut-off points. We noted the occurrence of two peaks of exposure on Sundays (10 a.m. and 4 p.m.) and three highest peaks reaching more than 40,000 per/h on Mondays (7 a.m., 1 p.m., and 5 p.m.), the latter corresponding mainly to professional and school mobility schedules (Figure 4.10). In parallel, the simulation of flood dynamics shows that the highest risk of flooding occurred between Sunday, September 8 at 9 p.m. and Monday, September 9 at 6 a.m. At this time, simulated traffic at potential cut-off points was the lowest. Finally, in order to better characterize the combined effect of these two dynamics, we constructed an index to quantify the number of motorists passing through the cut-off points at the time of their submersion, by weighting this number by the probability of submergence of each of the points crossed. The curve entitled "Persons potentially in danger," summarizing the dynamics of this exposure to danger, shows that the most critical time occurred during the peak traffic on Monday morning, between 7 and 9 a.m. (Figure 4.10). On average for the event, there were 15 motorists in danger per potentially submerged cut-off point. However, the cut-off point, where the fatal accident occurred in 2002, had 162 motorists passing through it, which is one of the maximum values in the study area.

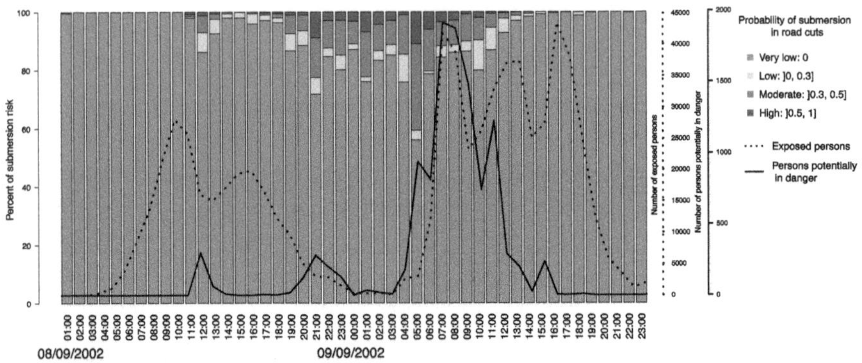

Figure 4.10. *Result of the simulation of the dynamics of motorists' exposure to road flooding in the Alès sector during the flash floods of September 8–9, 2002. Sources: [SHA 16]*

We also examined whether this exposure was evenly distributed according to the socio-demographic characteristics of individuals. Thus, we defined personal exposure (Pe) as the probability that a motorist will cross a submerged cut-off point on the routes traveled during the period of the event. To calculate this exposure, it was, therefore, necessary first to estimate the probability that not all the cut-off points crossed by an individual were submerged. The additional probability then provided the measure of personal exposure. Based on these calculations, we demonstrated that 75% of motorists had zero exposure during this event.

The average individual exposure was 0.17, but this varied significantly according to socio-demographic characteristics. Professional status (separating workers from other categories) was the most discriminating variable followed by profession and age, with gender being the last to be considered. Thus, the working population, whether technicians, intermediate occupations, workers, and managers, were the most highly exposed (0.27 for women and 0.33 for men). Conversely, inactive individuals were the least exposed (0.11 on average). Finally, students and working people in the professions of craftsmen, traders, farmers, and employees were moderately exposed (Av. Pe = 0.24).

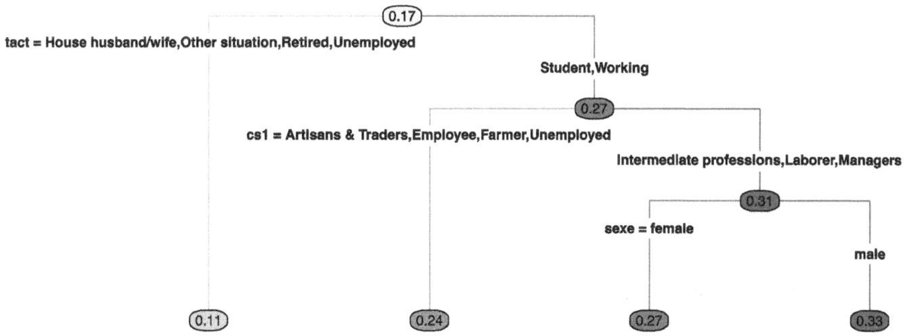

Figure 4.11. *Classification of individuals according to the effects of socio-demographic variables on exposure level. Source: [SHA 16]*

4.4.2.2. *The influence of attitudes, risk assessment on behavioral choices, and personal exposure*

MobRISK was then applied to the same case study, but instead of considering that individuals did not change anything about their daily travel activities, the decision model was implemented to take into account the effects of three types of behavioral adaptations related to the perception of environmental changes. Decisions to return home ("HOME") or stop and wait 60 minutes in front of a submerged road ("STOP") could be made when individuals were confronted with a submerged road on their route. "HOME" and 15-minutes delay of travel time ("DELAY") decisions were the choices available to individuals who completed the previous activity and prepared for a trip. Several scenarios were tested by varying standard attitudes and risk assessment separately or jointly.

– *Four standard attitude scenarios* define the conditional probability table assigned to the different behavioral decisions: (i) *attitude of non-adaptation* of behaviors, (ii) attitude of risk-taking (very low probabilities assigned to adaptation decisions), (iii) attitude of neutrality (equal probability between adaptation and non-adaptation decisions), and (iv) risk aversion (high probabilities assigned to adaptation decisions).

– *Three risk assessment scenarios* were tested: (i) an *underestimation of the risk meant that the* perception of the risk was always low regardless of environmental signs, (ii) an *"objective" risk assessment* distributed conditional probabilities according to visibility and road flooding levels, and (iii) an *overestimation of the risk* considering that the perception of the risk was always high regardless of environmental signs.

At the end of the multiple simulations testing the different combinations of scenarios, we can summarize the main results as follows:

– The adoption of decision rules corresponding to an attitude of risk aversion reduced the average exposure of motorists by half and reduced the percentage of people exposed by a factor of 3.

– Overestimation of the risk reduced motorists' exposure by half compared to a moderate estimate.

– The adoption of an attitude of risk aversion with an overestimation of the danger made it possible to completely eliminate motorists' exposure.

– Decisions to stop in front of a submerged road or to delay travel by re-planning daily activities reduced motorists' average exposure much more than returning home.

The model's performance during these first simulation tests was encouraging, although improvements are still to be expected in the coming years. The results obtained show that the timing of the event, as well as the choices of behavioral adaptation, are significant factors conditioning the average exposure level of motorists during flash floods. At the individual level, socio-demographic characteristics are also good predictors of mobility and, therefore, of exposure to road flooding.

4.5. Conclusion and perspectives

This chapter describes the modeling approach and results from the simulation of the MobRISK model for the flash flood event of September 8–9, 2002 in the Gard. Developed to assess the spatial and temporal dynamics of motorists' exposure to road flooding, this model is particularly useful for understanding the human risk that, in the case of flash floods, mainly threatens the people on the move who are the first victims.

The MobRISK simulator is, to our knowledge, the first to assess, on a French department scale, the dynamic exposure of populations by taking into account the variability of three components: (i) hydrometeorological circumstances and the probability of road flooding, (ii) daily mobility, and (iii) behavioral responses to danger. Other examples of micro-simulations generally use multi-agent modeling platforms (see for example [DAW 11]) to consider the dynamics of such interactions. However, these models are not developed at such large spatial scales (e.g. 6,000 km^2) and/or do not involve as many agents (about 700,000 inhabitants). The high temporal and spatial resolution of the model (in the order of magnitude of 1 minute and 1 meter) allows us to explore how flash floods disrupt and threaten daily activities and movements at the individual scale.

A first application of the model on the case study of the floods of September 8–9, 2002 in the Alès sector made it possible to implement an index (Pe) to quantify the exposure of more than 110,000 individuals on the basis of their movements corresponding to the event days studied. On this basis, we were able to show that the exposure index varied significantly according to occupational status and activity, age and gender, with men at a prime age being the most exposed. This study was also useful for locating the most dangerous road/river intersections in space and time because of the number of individuals likely to cross them when submerged. For this event, the temporal analysis of the model outputs indicated that the risk of flooding was highest during the night from Sunday to Monday, when traffic was expected to be the lowest. The simulation combining the risk of road flooding and the daily mobility expected for these weekdays confirmed a healthy temporal difference between the peak of cuts and the peak of mobility in this sector. Finally, on the basis of behavioral scenarios, we were able to test the combined effect of different standard attitudes to risk, risk assessment methods, and decision-making choices on individual exposure at the event level.

This first implementation showed the potential of such a tool for crisis management and emergency and road management planning. Nevertheless, the model can be improved by refining the modeled cognitive processes but also to optimize the calculation time and make it more operational. The next steps in its development are to better reproduce the travel times observed in the ENTD data. In fact, activity-centric modeling requires data on the location of individuals' activities. Within the framework of MobRISK, we only had access at the time to the location, at the communal resolution, of workplaces and places of study. To locate secondary activities, we used simple hypotheses of the proximity of these places of activity to the municipality of residence, which led us to underestimate the distances traveled for this type of activity and, therefore, to underestimate individual exposure overall.

Another possible source of improvement is to better take into account the link between exposure and human impact. Personal exposure is simply defined here as the probability of encountering a flooded road without regard to water level and current speed. This simplification is necessary due to the lack of precise data on the level of protection and morphology associated with road/river intersections. These aspects can be better taken into account by assimilating human impact data that are precisely located in space and time. This could include insurance data (road accidents during flood periods), emergency operations records, or social media data.

4.6. References

[AND 16] ANDRÉ-POYAUD I., CHARDONNEL S., CHARLEUX L. *et al.*, "Mobility and activity patterns of individuals and parenting couples in the metropolitan area of Grenoble (France)", available at: halshs.archives-ouvertes.fr, 2016.

[ANQ 10] ANQUETIN S., BRAUD I., VANNIER O. *et al.*, "Sensitivity of the hydrological response to the variability of rainfall fields and soils for the Gard 2002 flash-flood event", *Journal of Hydrology*, vol. 394, nos 1–2, pp. 134–147, 2010.

[ANT 01] ANTOINE J.-M., DESAILLY B., GAZELLE F., "Les crues meurtrières, du Roussillon aux Cévennes", *Annales de Géographie*, vol. 622, pp. 597–623, 2001.

[ASH 08] ASHLEY S.T., ASHLEY W.S., "Flood fatalities in the United States", *Journal of Applied Meteorology and Climatology*, vol. 47, pp. 805–818, 2008.

[BOT 09] BOTZEN W., AERTS J., VAN DEN BERGH J., "Dependence of flood risk perceptions on socioeconomic and objective risk factors", *Water Resources Research*, vol. 45, no. 10, p. W10440, 2009.

[BRA 10] BRAUD I., ROUX H., ANQUETIN S. *et al.*, "The use of distributed hydrological models for the Gard 2002 flash flood event: analysis of associated hydrological processes", *Journal of Hydrology*, vol. 394, no. 1, pp. 162–181, 2010.

[COO 13] COOLS M., CREEMERS L., "The dual role of weather forecasts on changes in activity-travel behavior", *Journal of Transport Geography*, vol. 28, pp. 167–175, 2013.

[CRE 13] CREUTIN J.D., BORGA M., GRUNTFEST E. *et al.*, "A space and time framework for analyzing human anticipation of flash floods", *Journal of Hydrology*, vol. 482, pp. 14–24, 2013.

[DAM 81] DAMM D., LERMAN S.R., "A theory of activity scheduling behavior", *Environment and Planning A*, vol. 13, no. 6, pp. 703–718, 1981.

[DAW 11] DAWSON R.J., PEPPE R., Wang M., "An agent-based model for risk-based flood incident management", *Natural Hazards*, vol. 59, no. 1 , pp. 167–189, 2011.

[DEB 16] DEBIONNE S., RUIN I., SHABOU S. *et al.*, "Assessment of commuters' daily exposure to flash flooding over the roads of the Gard region, France", *Journal of Hydrology*, vol. 541, pp. 636–648, 2016.

[DIA 13] DIAKAKIS M., DELIGIANNAKIS G., "Vehicle-related flood fatalities in Greece", *Environmental Hazards*, vol. 12, pp. 278–290, 2013.

[DIJ 59] DIJKSTRA E.W., "A note on two problems in connexion with graphs", *Numerische Mathematik*, vol. 1, no. 1, pp. 269–271, 1959.

[ETT 09] ETTEMA D., VAN DER LIPPE T., "Weekly rhythms in task and time allocation of households", *Transportation*, vol. 36, no. 2, pp. 113–129, 2009.

[FOS 16] FOSSET P., BANOS A., BECK E. *et al.*, "Exploring intra-urban accessibility and impacts of pollution policies with an agent-based simulation platform: GAMIROD", *Systems*, vol. 4, no. 1, p. 5, 2016.

[FRE 83] FRENCH J., ING R., VON ALLMEN S. *et al.*, "Mortality from flash floods: a review of national weather service reports, 1969–81", *Public Health Reports*, vol. 98, pp. 584–588, 1983.

[HAN 81] HANSON S., HANSON P., "The travel-activity patterns of urban residents: dimensions and relationships to sociodemographic characteristics", *Economic Geography*, vol. 57, no. 4, pp. 332–347, 1981.

[JON 05] JONKMAN S.N., KELMAN I., "An analysis of the causes and circumstances of flood disaster deaths", *Disasters*, vol. 29, pp. 75–97, 2005.

[KIL 07] KILPELÄINEN M., SUMMALA H., "Effects of weather and weather forecasts on driver behaviour", *Transportation Research Part F: Traffic Psychology and Behaviour*, vol. 10, no. 4, pp. 288–299, 2007.

[LIN 08] LINDELL M.K., HWANG S.N., "Households' perceived personal risk and responses in a multihazard environment", *Risk Analysis*, vol. 28, no. 2, pp. 539–556, 2008.

[LIN 12] LINDELL M.K., PERRY R.W., "The protective action decision model: theoretical modifications and additional evidence", *Risk Analysis*, vol. 32, no. 4, pp. 616–632, 2012.

[LU 99] LU X., PAS E.I., "Socio-demographics, activity participation and travel behavior", *Transportation Research Part A: Policy and Practice*, vol. 33, no. 1, pp. 1–18, 1999.

[MAP 09] MAPLES L.Z., TIEFENBACHER J.P., "Landscape, development, technology and drivers: the geography of drownings associated with automobiles in Texas floods, 1950–2004", *Applied Geography*, vol. 29, pp. 224–234, 2009.

[NAU 13] NAULIN J.-P., PAYRASTRE O., GAUME E., "Spatially distributed flood forecasting in flash flood prone areas: application to road network supervision in southern France", *Journal of Hydrology*, vol. 486, pp. 88–99, 2013.

[RUI 07] RUIN I., Conduite à contre-courant. Les pratiques de mobilité dans le Gard: facteur de vulnérabilité aux crues rapides, PhD thesis, Université Joseph-Fourier-Grenoble I, Grenoble, France, 2007.

[RUI 08] RUIN I., CREUTIN J.-D., ANQUETIN S. *et al.*, "Human exposure to flash floods – relation between flood parameters and human vulnerability during a storm of September 2002 in Southern France", *Journal of Hydrology*, vol. 361, pp. 199–213, 2008.

[RUI 14] RUIN I., LUTOFF C., BOUDEVILLAIN B. *et al.*, "Social and hydrological responses to extreme precipitations: an interdisciplinary strategy for postflood investigation", *Weather, Climate and Society*, vol. 6, no. 1, pp. 135–153, 2014.

[SHA 12] SHARIF H.O., HOSSAIN M.M., JACKSON T. *et al.*, "Person-place-time analysis of vehicle fatalities caused by flash floods in Texas", *Geomatics, Natural Hazards and Risk*, vol. 3, pp. 311–323, 2012.

[SHA 16] SHABOU S., Extrême météorologiques & exposition sur les routes. Contribution à MobRISK: modèle de simulation de l'exposition des mobilités quotidiennes aux crues rapides, PhD thesis, Université Grenoble Alpes, Grenoble, France, 2016.

[SHA 17] SHABOU S., RUIN I., LUTOFF C. *et al.*, "MobRISK: a model for assessing the exposure of road users to flash flood events", *Natural Hazards and Earth System Sciences.*, vol. 17, pp. 1631–1651, 2017.

[SPI 14] ŠPITALAR M., GOURLEY J.J., LUTOFF C. *et al.*, "Analysis of flash flood parameters and human impacts in the US from 2006 to 2012", *Journal of Hydrology*, vol. 519, pp. 863–870, 2014.

[TER 15] TERTI G., RUIN I., ANQUETIN S. *et al.*, "Dynamic vulnerability factors for impact-based flash flood prediction", *Natural Hazards*, vol. 79, no. 3, pp. 1481–1497, 2015.

[TER 16] TERTI G., RUIN I., ANQUETIN S. *et al.*, "A situation-based analysis of flash flood fatalities in the United States", *Bulletin of the American Meteorological Society*, vol. 98, no. 2, pp. 333–345, 2016.

[VAN 16] VANNIER O., ANQUETIN S., BRAUD I., "Investigating the role of geology in the hydrological response of Mediterranean catchments prone to flash-floods: regional modelling and process understanding", *Journal of Hydrology*, vol. 541, pp. 158–172, 2016.

[VER 10] VERSINI P.-A., GAUME E., ANDRIEU H., "Assessment of the susceptibility of roads to flooding based on geographical information–test in a flash flood prone area (the Gard region, France)", *Natural Hazards and Earth System Science*, vol. 10, no. 4, pp. 793–803, 2010.

5

Assigning Travel-Activity Patterns Based On Socio-Demographics for Flood Risk Assessment[1]

5.1. Introduction

The increasing frequency of natural disasters is one of the major implications of climate change. Due to their rapidness and suddenness, flash floods are among the most dangerous natural hazards causing an important number of victims. Several studies showed that over half of flash flood victims are motorists trapped by road flooding on their daily route [ASH 08], [SHA 12], [TER 16]. Therefore, flash flood risk assessment studies have become increasingly interested in measuring the impact of flash flooding on road network and mobility behaviors. The Gard region, located in southern France, is frequently exposed to heavy rainfall events leading to severe and highly damageable flash flood events. Antoine *et al.* [ANT 01] recorded 27 fatal flood events causing 277 deaths in this flood-prone region between 1316 and 1999. Almost half of these victims were motorists trapped in their vehicles while driving on flooded roads. In order to address road flooding risk in the Gard region, Versini *et al.* [VER 10] and Naulin *et al.* [NAU 13] proposed a tool enabling the prediction of road submersion based on an inventory of road submersions over the last 40 years. This tool consists in

Chapter written by Saif SHABOU, Isabelle RUIN, Céline LUTOFF, Sonia CHARDONNEL and Samuel DEBIONNE.

1 The work developed in this chapter was funded by the French National Research Agency (ANR) through the project MobiCLIMeX ANR-12-SENV-0002-01 and is part of the MISTRAL-HyMeX framework.

combining road sensitivity to flooding (based on geomorphologic characteristics) and simulated water discharge in order to provide an estimation of submersion probability in road/river intersections (called "potential road cuts"). More recently, Debionne *et al.* [DEB 16] proposed a model, called *MobRISK*, for assessing commuters' exposure to flash flooding over the roads in the Gard region. A preliminary application of *MobRISK* consisted in simulating commuting routes of the Gard population and quantifying their exposure based on the number of potential road cuts that each individual would cross on his way (see Chapter 4 of this book). The main results indicate that commuters who have longer itineraries are the most exposed to road flooding. The analysis of socio-economic profiles of exposed motorists showed that commuters who are residents of rural municipalities, as well as professionals with highly qualified jobs, are the most exposed persons. This first application of *MobRISK* model displayed a trip-based approach for travel demand modeling and only cumulated daily exposure is computed without taking into account the temporal dynamics of population mobility and road submersions. In addition, only commuting trips are simulated without considering the variety of travel motivations, such as leisure and shopping activities, which might influence one's willingness to pursue his journey regardless of the degradation in travel's conditions. These limitations are essentially due to the only use of census data, which do not provide the necessary information for simulating travel-activity patterns. The next step for improving motorists' exposure assessment in *MobRISK* model consists in including the temporal dynamics of travel behavior and various travel motivations in order to fully characterize spatio-temporal motorists' exposure [DEB 16]. This chapter proposes a method for assigning travel-activity patterns to the Gard region population in order to address these limitations in the current version of *MobRISK*.

Advances in transportation research have shown the emergence of a new paradigm in travel demand modeling represented by the activity-based approach. These mobility models enable the simulation of travel-activity behaviors at the household and individual levels. One of the main challenges represented by the activity-based mobility models consists in simulating individual activity schedules at a high temporal resolution. Various statistical and computational schedule simulation methods have been developed depending usually on diary data availability and accuracy [ARE 00], [AUL 09], [ETT 93]. Generally, socio-demographic characteristics and household composition description are used as explanatory variables for predicting travels' number and activity participation. Although schedule simulators enable the identification of activity scheduling rules, the simulated travel-activity patterns are hardly able to generate the complexity of individual activity scheduling process. To preserve the internal diversity of activity schedules, we propose in this paper a method for assigning real observed activity schedules to individuals based on their socio-demographic description. Several studies examined the interaction between personal and household constraints and mobility behavior [AND 16], [PAS 84], [VAU 99]. However, much less attention is given to the sequential nature of travel-activity patterns. Application of Sequence Analysis methods in social science, particularly in life course studies, illustrated the benefits of using such methods for better characterizing, comparing, and classifying sequences-like objects. In addition, Studer *et al.* [STU 11] proposed an original method to assess the effect of categorical variables on dissimilarities between activity sequence patterns. We propose in this study to test this method for analyzing activity schedules in order to extract rules of activity schedules assignment based on socio-demographic variables. Those rules could be used then for improving the so-called *MobRISK* model by integrating temporal dynamics in mobility simulations. Therefore, we use travel-activity data, provided by the French National Travel Activity survey, corresponding to the study region selected in the *MobRISK* model [DEB 16].

This chapter is organized as follows. In section 5.2, we present the basic concepts of the activity-based approach in mobility behavior modeling and a literature review of the relationship between socio-demographic characteristics and travel-activity behavior. Then, the benefits of Sequence Analysis methods in exploring travel-activity sequences will be discussed based on previous applications. In section 5.3, we describe the data used for this study and sequence analysis methodology used in order to assign travel-activity sequences to individuals based on their socio-demographics. Section 5.4 presents the results of schedule assignment model and the evaluation of the performance of the assignment. Finally, we discuss the limitations of the proposed methods and we present some conclusions and directions for future work.

5.2. Travel-activity behaviors

This section provides a review of basic concepts used in activity-based travel demand modeling and discusses the benefits of integrating sequence analysis methods for quantifying the effects of socio-demographic variables on travel-activity behavior.

5.2.1. *Travel demand models*

The main goal of analyzing travel-activity behaviors in this study consists in assessing the exposure of road users to road flooding during extreme hydro-meteorological events. Some studies highlighted the importance of travel motivation and trip purposes on travel adaptation behavior in weather disruption. For instance, Cools *et al.* [COO 10] demonstrated that travel change decisions with respect to weather conditions depend on trip purposes and that leisure and shopping activities are more susceptible to be canceled and postponed than work/school activities. By analyzing reported behaviors based on a post-flood investigation, Ruin *et al.* [RUI 14] showed the important role of perceived importance and flexibility of planned activities in mobility adaptation capacities during past flash flood events. These findings highlight the relevance of considering travel motivations on travel demand simulation to better assess mobility exposure to heavy precipitation and road flooding.

Recent advances in travel demand modeling shifted from a trip-based approach that focuses on the quantification of the trips generated by population mobility to an activity-based approach. Activity-based models emerged in the 1990s in reaction to the recognition of the importance of trips' sequences and travel motivations for a better understanding of population mobility behavior. They offer an appropriate framework to micro-simulate individual travel-activity patterns [RAS 14], [TIM 02]. The theoretical foundation of activity-based paradigm is based on the seminal work of Hägerstand [HAG 70] and Chapin [CHA 74] in Time Geography, which emphasizes the role of personal characteristics and spatial-temporal constraints in travel-activity behavior. According to the activity-based approach, travel behavior is considered as derived from the need to pursue different activities spatially and temporally distributed [BEN 98], [BHA 99a], [BOW 01]. Consequently, activity-based models aim at predicting how, why, when, how often, where, and with whom the different activities are conducted by the individuals. This approach focuses more on activity sequences and chains than a simple counting of trips. McNally [MCN 95] identified the most important specificities in activity-based modeling: (i) travel is derived from the demand for activity participation, (ii) sequences and patterns of travel behavior are the units of analysis instead of individual trips in trip-based models, (iii) household and socio-demographic characteristics affect travel-activity behavior, and (iv) spatial, temporal, and interpersonal factors that constrain activity-travel patterns are taken into account. Over the past years, several activity-based models have been developed: TRANSIMS [BOW 97], ALBATROSS [ARE 00], CEMDAP [BHA 04], MATSIM [BAL 06] ADAPTS [AUL 09], and MIRO [FOS 16]. The main differences between these models consist in the spatial–temporal resolution of travel simulations and the methodology used for generating individual schedules and travel-activity patterns. Two main approaches have been suggested for modeling individual activity schedules: utility-maximizing models and rule-based models. The utility-maximizing models are based on the microeconomic theory. They assume that individuals make rational behavioral choices by attributing various utility values to their daily activities and attempt to maximize their daily travel-activity utilities [REC 86]. However, these models make unrealistic assumptions about the rationality of individuals' scheduling behavior. Inspired by psychological studies and decision process theories, the so-called "rule-based" models aim at introducing decision rules and heuristics to represent activity scheduling processes [GAR 94], [GAR 98], [GOL 94]. The definition of these rules is usually based on a statistical analysis of individual travel-activity data and time-use surveys by evaluating the relationships between socio-demographic characteristics and travel-activity behavior. Recent efforts in activity-based travel-demand modeling combine the utility-maximizing assumptions with the rule-based approach for simulating travel-activity behavior based on socio-demographic profiles [FOS 16].

Several studies have been conducted to determine the effect of individual socio-demographic characteristics and household composition on activity participation and daily travel-activity behaviors [HAN 81]. Lu and Pas [LU 99] showed that the variables that are systematically important for explaining variations in activity participation include age, gender, employment status, and number of children. Pas [PAS 84] and Elias *et al.* [ELI 08] found that gender is the most significant factor for explaining daily travel-activity behavior. Bhat and Misra [BHA 99b] focused on the effect of age on travel behaviors and demonstrated a decrease in activity space with increasing age. Perchoux *et al.* [PER 14] indicated that age, gender, employment status, and education level were strong determinants of individual spatial behavior regarding the size of the activity space, the centering of activity space on the residential neighborhood, and the volume of activities. In terms of activity participation, Ettema and van der Lippe [ETT 09] have highlighted that gender, marital status, and the presence of children significantly affect travel number and timing of leisure activities. André-Poyaud *et al.* [AND 16] performed a classification analysis of travel-activity data provided by French National Household Transport Survey. Eight types of individual mobility were identified, depending essentially on gender, lifecycle, age, employment status, and work places. Vaughn *et al.* [VAU 99] applied a multivariate regression tree procedure to identify socio-demographic variables that are relevant in explaining the variation in household activity-travel patterns. The results show that the number of workers of school-age children and the marital status better explain travel's behaviors related to work, school, and in-home activities than travels related to maintenance and recreation activities. The regression tree results were used for assigning household travel-activity patterns from observed patterns based on household travel survey to synthetic households in the TRANSIMS mobility model development [BAR 95], [SMI 95].

Although the mentioned studies examined essentially the effects of socio-demographic characteristics based on travel variables such as travel frequency and activity participation and duration, they do not take into consideration the sequential nature of activity patterns. In order to fully characterize individual travel-activity patterns, the sequential dimension should be included in the analysis. Recently, some authors highlighted the benefit of using sequences analysis methods, widely used in molecular biology, for exploring travel-activity behaviors [JOH 01], [JOH 11], [LES 14]. In the next subsection, we will present the advantages of sequences analysis methods and its interest for our own research objectives.

5.2.2. *The advantages of using sequences analysis methods*

Sequence analysis methods were first developed in the field of coding theory and string editing to detect dissimilarities between long strings of codes. These methods have been then widely used in molecular biology and bioinformatics since they provide a rigorous framework for analyzing and comparing DNA, RNA, or protein sequences by identifying regions of similarities [KRU 83]. Thanks to the benefits of sequence analysis techniques in extracting rich knowledge from a sequence like objects, they have gained increased attention in social science, particularly in demographic and life courses studies. Abbott and Forrest [ABB 86] introduced these methods for analyzing professional careers and familial life course, where individual trajectories are considered as a series of states evolving with time. The popularity of sequence analysis techniques in life course studies is largely due to the holistic approach they provide by considering a complete life sequence as a single unit of analysis [BIL 06]. Blanchar and Fillieule [BLA 11] pointed out five major objectives of sequence analysis techniques. First, they enable the description and the visualization of groups of sequences. They are also often used to perform comparison and classification operations. Third, several sequences' analysis tools are developed for sequence data mining and extracting rich knowledge and insights from hidden patterns. Finally, they can be used to identify causal relationships between external variables and sequences characteristics.

In transportation research, sequences analysis methods are mainly used for measuring the similarity among travel-activity patterns and identifying clusters of mobility behaviors. A second use of these methods in the travel-activity analysis consists in measuring the similarities between predicted and observed travel patterns to quantify the goodness of fit in travel predictive models. Wilson [WIL 98] applied sequences comparison tools in time-use diary data to examine the timing of activities participation and compare individual diaries. Lesnard and Kan [LES 09] proposed an advanced sequences alignment metric for the analysis of UK time-use survey. Various typologies of workdays and workweeks were retrieved, exhibiting different work activity timing and organization. Joh *et al.* [JOH 01] developed a multidimensional sequence alignment method to compare activity-travel patterns by integrating different dimensions (e.g. activity location, transport modes, activity types, etc.). Joh and Timmermans [JOH 11] applied sequences analysis methods to perform classification analysis (based on decision tree procedure) of activity-travel diary data and identify various cluster profiles together with their representative sequence. In order to optimize multidimensional sequence alignment, Kwan *et al.* [KWA 14] proposed a multi-objective optimization evolutionary algorithm for better assessing activity pattern similarities. The mentioned studies highlighted the benefits of using sequence analysis methods for better characterizing daily travel-activity patterns.

The sequential character of activity scheduling seems to be essential to understand mobility adaptation behavior in crisis circumstances, such as flood events. Ruin *et al.* [RUI 14] conducted a post-flash flood event investigation, in southern France, enabling to reproduce activity patterns of surveyed persons regarding the hydro-meteorological dynamics. Collected data showed that individuals have a tendency to implement a sequence of protection activities (Information-Organization-Protection) with respect to their daily activity patterns. In the next section, we describe the travel-activity data used for this study. Then, we detail the methodology used for sequences comparison and the assessment of the effect of categorical variables on sequences dissimilarities developed by Studer *et al.* [STU 11].

5.3. Data and methods

5.3.1. *Data*

The travel-activity data used in this study originate from the French National Transport and Travel Survey (ENTD) carried out by the INSEE (French National Institute of Statistics and Economic Studies) from 2007 to 2008. A sample of 20,220 households were surveyed in Metropolitan France: 10,700 households concern the national sample and the rest cover regional extension carried out in the following regions: Ile-de-France, Pays de Loire, Bretagne, Languedoc-Roussillon, and Midi-Pyrénées. The responders were requested to indicate their socio-demographic characteristics (age, gender, professional status, etc.), their household composition, and their mobility description during one weekday and one weekend before the visit day. They were instructed to mention the different trips they made during the surveyed days (from 4 a.m. to 4 a.m. of the following day), transport modes, trips' purposes, time of departure and arrival, and traveled distance. A coding scheme including 10 categories of activity was used for the survey: working, studying, administrative procedure, medical treatment, accompanying, shopping, visiting, holidays, leisure, and home activities.

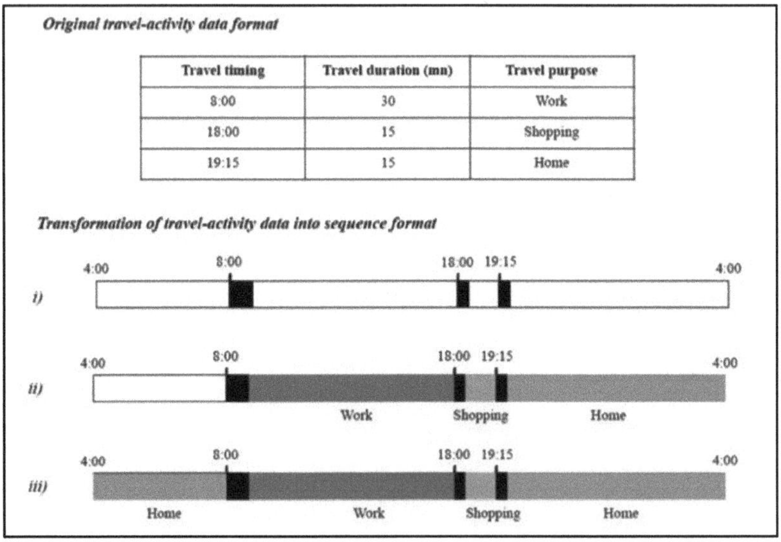

Figure 5.1. *Schematic example representing the necessary steps for transforming original travel-activity data format (travels' table) into activity sequence format. For a color version of this figure, see www.iste.co.uk/lutoff/hydrometeo2.zip*

The original data are provided in a trip-based form representing a recording of every individual travel, departure and arrival time, and travel purpose. We transformed the original travel data into activity sequences format within 5-minute time step. The method follows three steps: (i) creating an empty schedule for every individual constituted of a sequence of 5-minute duration activities; (ii) filling the empty schedules with travel activities with respect to the mentioned time of departure and arrivals; and (iii) completing the periods in between travels with the trip purposes with respect to activity durations. Figure 5.1 illustrates a schematic representation of the steps used for transforming the original travel-activity data format into activities' sequence.

5.3.2. *Activity sequences comparison*

The methods used for comparing a pair of sequences consist in measuring the dissimilarities between them as a function of the number and type of operations needed to transform one sequence into the other (to match them). The operations considered are insertions/deletions or substitutions of elements. Figure 5.2 illustrates a matching of a pair of sequences in two different ways regarding the type of operation: (i) using only substitutions by replacing the different elements of one sequence by those in the second one and (ii) using a combination of insertions and deletions operations. In sequence analysis literature, we can distinguish three distance metrics: Hamming distance [HAM 50] and two versions of Levenshtein distance [LEV 66]. The difference between these distance metrics consists in the type of operations used for matching the sequences. The Hamming distance is based only on the number of substitutions needed to match two sequences. The basic constraint of this method is that the sequences have to be of equal length. Levenshtein [LEV 66] suggested using insertions and deletions operations to measure the distance between sequences, which facilitate the detection of local similarities between sequences with different length. He proposed two distance metrics: the first – called Optimal Matching method – consists in taking into account both operations (insertion/deletion and substitution) and the second focuses only on the insertion/deletion operation (Table 5.1[2]).

2 Cf. Annex 2.

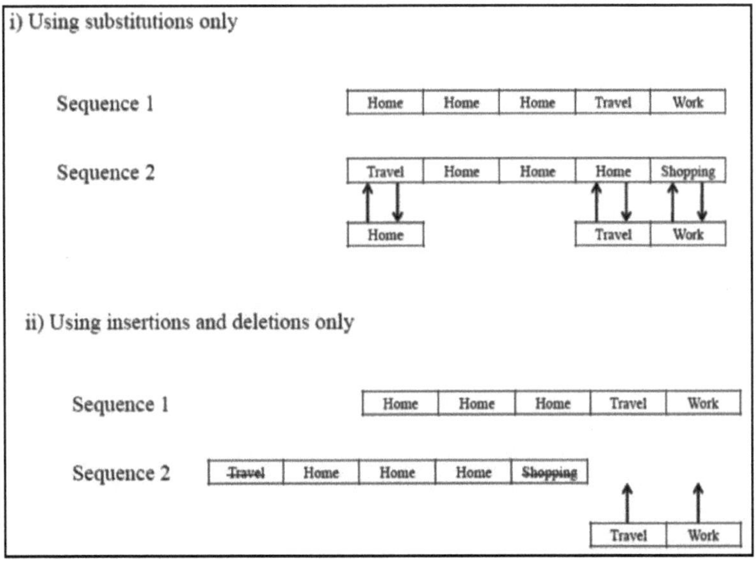

Figure 5.2. *Schematic representation of sequence matching operations*

In a social science application of sequence analysis techniques, Optimal Matching (OM) metric is often preferred because it offers more flexibility and adaptability in sequence comparison [LES 09]. Since the distances between a pair of sequences are based on the number of operations needed to match them, OM methods offer the possibility to assign different costs to insertions/deletions and substitutions operations. These costs consist in attributing weights to the different operations in order to customize distance computation. The costs' definitions constitute a major methodological problem in OM applications [GAU 09]. The classical way to use OM metric is to attribute a cost equal to one for both substitutions and insertion/deletion operations. When all the substitution costs are strictly higher than insertions/deletions costs, or only insertions/deletions operations are performed, the distance computed represents the longest common subsequences independently of their location in the sequence [KRU 83]. On the contrary, when substitution costs are lower than insertions/deletions, or only substitution operations are allowed, the distance will focus more on the contemporary similarities between the sequences [LES 14]. The second option of customizing the OM metric consists in giving different costs/weights to the substitution operations themselves by assigning different distances between the states of the sequences according to their similarities in one considered dimension. For instance, given a pair of activity sequences to be compared, leisure and shopping activities can be considered closer than leisure and work activities in term of timing flexibility. These similarities can be taken into account by giving lower costs of substituting a leisure activity by shopping than working.

5.3.3. *Discrepancy analysis*

In addition to sequences similarity measurement, some sequences analysis methods enable the assessment of the effects of different variables on sequences variability. Studer *et al.* [STU 11] introduced a discrepancy analysis method that enables the measurement of the relationships between categorical variables and a set of sequences described by a matrix of dissimilarities. The proposed method consists in an adaptation of ANOVA statistical test for a sequence like objects. The fundamental logic of classic ANOVA test is to group individuals based on one or more factors and compare the inter-group and intra-group variances in order to measure how much the chosen factors can explain the total variance. In the ANOVA test, the variance is measured using the "the sum of squares" (SS), which is the sum of the squared Euclidian distances between each value and the mean. In order to apply the ANOVA test to complex objects, such as sequence like objects, Anderson [AND 01] proposed to express the sum of squares using the average pairwise squared Euclidian distances ($d^2_{e,i,j}$) and then replace the Euclidian distance ($d_{e,i,j}$) with any non-Euclidian measure of dissimilarity ($d_{i,j}$) (equation [5.1])

$$SS = \sum_{i=1}^{n}(y_i - \bar{y})^2 = \frac{1}{2n}\sum_{i=1}^{n}\sum_{j=1}^{n}(y_i - y_j)^2 = \frac{1}{n}\sum_{i=1}^{n}\sum_{j=i+1}^{n}d_{ij}^2 \qquad [5.1]$$

The principle of ANOVA test consists in considering the total variance (SS_{Total}) as a sum of between-group variance, so the variance explained by a covariate $SS_{Explained}$, and a residual within-group variance ($SS_{Residual}$) (equation [5.2]) [BAT 88]. Hence, we can compute the part of discrepancy explained by a categorical variable (R^2) and compare the explained discrepancy with the residual discrepancy (equations [5.2] and [5.3]).

$$SS_{Total} = SS_{Explained} + SS_{Residual} \qquad [5.2]$$

$$R^2 = \frac{SS_{Explained}}{SS_{Total}} \qquad [5.3]$$

$$F_{observed} = \frac{SS_{Explained}/(m-1)}{SS_{\text{Re}\,sidual}/(n-m)} \qquad [5.4]$$

where m refers to the number of groups and n is equal to the sample size.

Studer *et al.* [STU 11] proposed the use of permutation tests in order to assess the statistical significance of the explained discrepancy by a covariate instead of the Fisher's F distribution classically employed in the ANOVA. The permutation test consists in changing the group of the considered covariate to which belong the sequence through a random permutation and computing the F statistic for every permutation ($F_{permuted}$). By repeating these operations n times, a distribution of $F_{permuted}$ is obtained. Thus, the proportions of $F_{observed}$ that are higher than $F_{permuted}$ reveal the significance of $F_{observed}$ [AND 01]. The number of permutations is generally fixed at 1,000 permutations for tests with 5% significance level and 5,000 permutations for tests with 1% significance level [MAN 06].

In addition to discrepancy analysis, Studer *et al.* [STU 11] proposed a complementary regression tree analysis, which enables to classify the sequences with respect to the results obtained from the discrepancy analysis. Regression trees are techniques used for modeling the relationship between dependent (e.g. individual activity schedule) and independent (e.g. socio-demographic characteristics) variables. The regression tree techniques are useful when there are several explanatory (independent) variables, and complex relationship between dependent and independent variables. They produce homogeneous groups of sequences regarding the dependent variable. The application of regression tree analysis in life trajectory data provided a classification of various life trajectories depending on socio-demographic variables [STU 11].

5.3.4. *Schedule assignment model evaluation*

The main objective of using the regression tree analysis in this study is to extract rules for assigning activity schedules to individuals based on their socio-demographic characteristics. One major limitation of using such data clustering methods relies on the difficulty to evaluate the obtained results. An evaluation method, classically used in this kind of studies, consists in partitioning the total data into *test* and *training* sets. The training data set, usually representing two-thirds of the total data, is used in order to build the classification model. Then, the test data set is used to evaluate the model.

The selection of the training data set is often implemented by random sampling from the total data. However, in our case travel-activity data, classification depends largely on the socio-demographic variables. Therefore, we propose a supervised sampling method by sampling two-thirds of the total population with respect to the socio-demographic description of the population (age, gender, household composition, professional status, and profession). This means that the socio-demographic description of the training population data and the total population data is similar (based on the listed variables). Once the classification rules are obtained from the training data, we apply them to the test set. Then, for every individual, we compute a similarity distance between his observed schedule and the assigned schedule obtained from the classification rules. A model performance index is then computed based on the difference between the observed/assigned schedules distance and an average distance between the schedule and all the other schedules. This performance index enables us to assess the capability of the used method for assigning schedules to individuals as similar as possible to their real schedules.

5.4. Results

This section presents some descriptive statistics of the travel-activity data corresponding to the Languedoc-Roussillon region and the results of the application of discrepancy analysis of activity sequences for quantifying the effect of socio-demographic variables in sequences' dissimilarities.

5.4.1. *Descriptive statistics*

The population sample is constituted of 1,240 individuals, representing 2,560,870 habitants of the Languedoc-Roussillon region. The socio-demographic description of the sample is provided in Table 5.2[3]. The average number of trips per individual is around 3.7 trips for an average weekday and 3.2 trips for an average weekend. Some socio-demographic variables seem to have an effect on the frequency of daily travels. During weekdays, women, young adults to middle-aged people (from 18 years old to 45 years old), unemployed, and individuals who are living in households with a larger number of persons and cars perform more travels. However, during weekend, men, working people, and couples living with children are

3 Ibid.

performing more trips (Table 5.3[4]). These differences can be explained by personal and household daily needs and constraints. When we examine trips' goals, we find that more than 40% of performed trips are home destination for both weekday and weekend. Logically, we observe that travels for school, work, medical, administrative, and accompanying activities are mostly performed during weekdays. On the other hand, travels for shopping, visiting, and leisure activities are rather performed during the weekend (Figure 5.3).

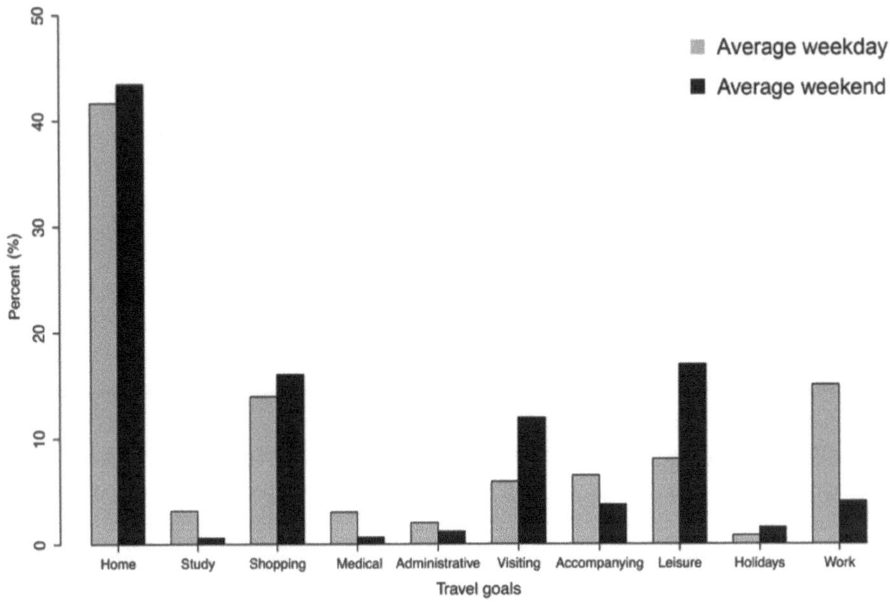

Figure 5.3. *Differences in travel goals between weekdays and weekend*

Concerning activities' durations, individuals spend in average 18 hours in their home place during an average weekday and 19 hours during an average weekend day. Not surprisingly, the main differences in time use between weekdays and weekends are related to work/school activities, which have longer durations in weekdays and secondary activities, which are longer in weekend (Figure 5.4).

4 Ibid.

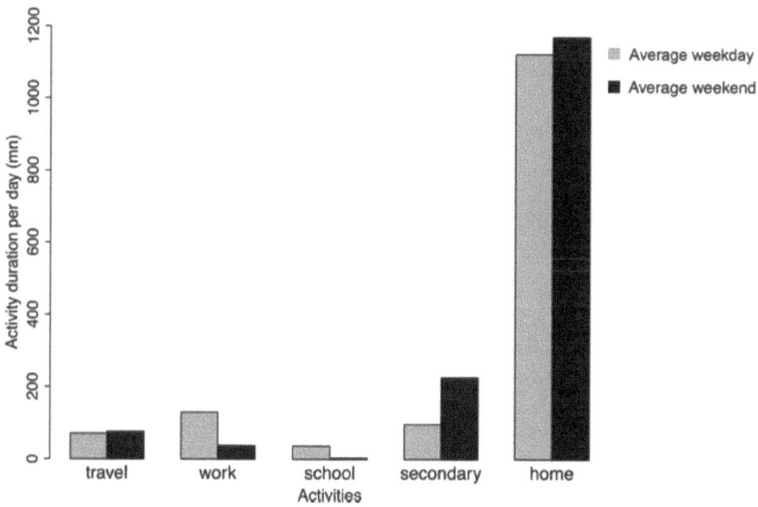

Figure 5.4. *Differences in activity durations between weekdays and weekend*

Furthermore, the socio-demographic characteristics of the population seem to have a significant effect on travel durations. In weekdays, men perform longer itineraries than women. Persons from 18 to 29 years old and especially students also have longer travel duration than unemployed and working people (Figure 5.5).

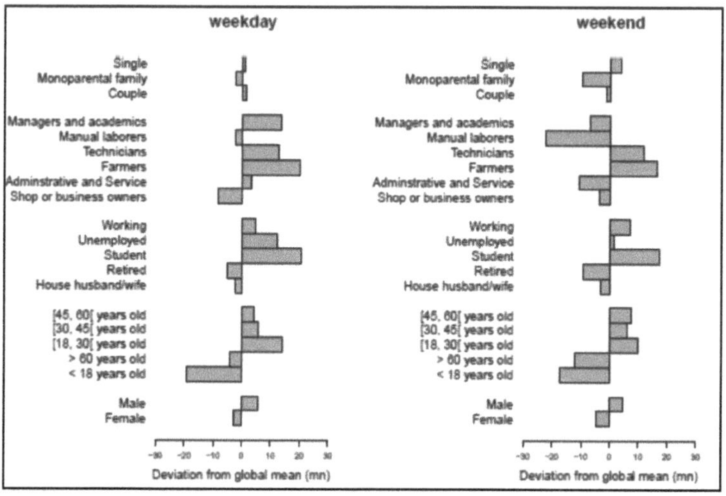

Figure 5.5. *Travel durations according to socio-demographic profiles in weekday and weekend*

The comparison of weekdays and weekend travels' duration shows that elderly people, unemployed, and some professions (managers, manual laborers, administrative, and services) spend much less time in traveling during weekends than the other categories.

5.4.2. *Schedule analysis*

Considering travel frequencies and activity durations is not sufficient to fully characterize travel-activity behaviors, especially for analyzing activity rescheduling decisions in reaction to weather disruption. Sequence analysis methods offer the possibility to identify hidden activity patterns and to explore the sequential dimension of mobility behavior. An individual schedule can be visualized as a sequence of activities representing, for every individual, the performed activity in every time step (Figure 5.6).

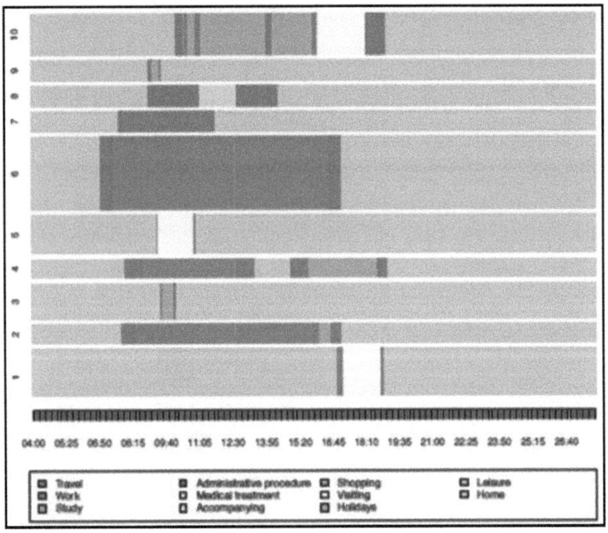

Figure 5.6. *A visualization of a sample of 10 activity sequences. The X-axis represents the time of day (5-minutes time step). Each horizontal line in the Y-axis represents an individual sequence of activities differentiated by the colors. The lines' width displays the statistical weights of the individuals regarding the survey data. For a color version of this figure, see www.iste.co.uk/lutoff/hydrometeo2.zip*

The temporal distribution of performed activities by the total population during an average weekday shows two main peaks at 10 a.m. and 4 p.m. corresponding to "work" and "school" activities timing (Figure 5.7a).

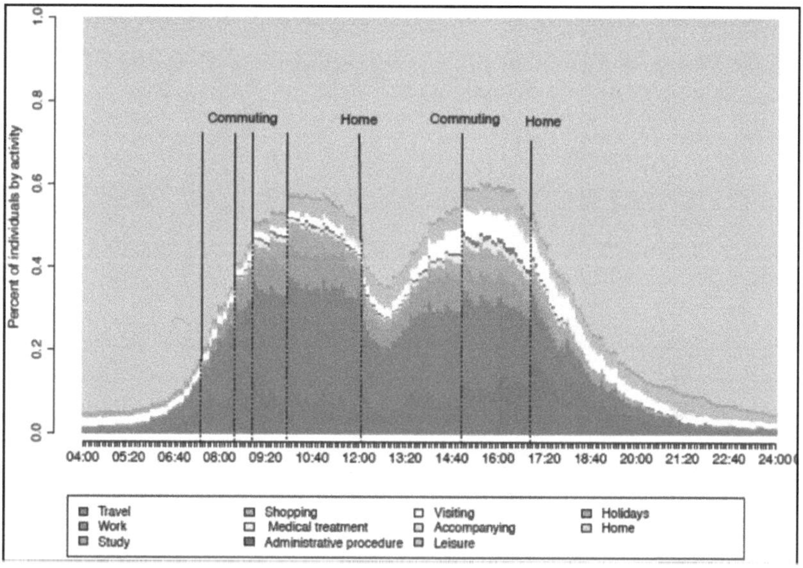

Figure 5.7a. *Activity schedules state distribution in an average weekday. The X-axis represents the time of day (5-minute time step). The Y-axis represents the percent of individuals performing the different activities in each time step*

We can identify four peaks of commuting during the day (from 7 a.m. to 10 a.m.) followed by trips to home for lunch (at 12 a.m.). In the afternoon, we can distinguish a peak of commuting travels at 3 p.m. and a peak of trips to return home at 5 p.m. On the other hand, the temporal distribution of weekend activities shows two main peaks of "leisure" and "visiting" activities at 11 a.m. and 4 p.m. (Figure 5.7b). These peaks are preceded by different waves of travels, especially at 9 and 10 a.m. for morning activities and 2 and 3 p.m. for afternoon activities.

As indicated in section 5.3.2, the use of OM metric for comparing sequences requires the definition of the costs attributed to activity substitutions. Two cost settings are tested in this study in relation with the category of activities: traveling, in home activities, out of home mandatory activities (work and school), and out of home secondary activities (shopping, administrative, medical, accompanying, and leisure activities). The first default cost setting maintains an equivalent substitution cost (c = 1), used for all substitutions operations (Table 5.4a[5]). Then, we proposed a customized

5 Ibid.

substitutions cost setting with respect to the activity category in such a way that lower costs are attributed to substitutions among activities belonging to the same category (mandatory versus secondary) (Table 5.4b[6]).

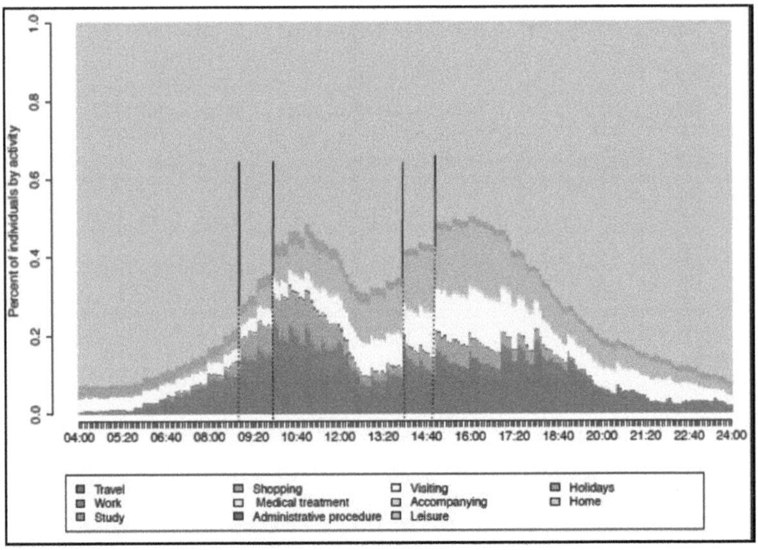

Figure 5.7b. *Activity schedules state distribution in an average weekend. The X-axis represents the time of day (5-minute time step). The Y-axis represents the percent of individuals performing the different activities in each time step*

We conducted a multifactor discrepancy analysis of activity sequences for an average weekday and an average weekend in order to assess the effects of eight selected variables (gender, age, education level, professional status, household size, number of children, household type, and number of cars) on schedule dissimilarities. The choice of these variables is based on previous studies on the effect of socio-demographic characteristics on daily travel-activity behavior [PAS 84]. The results show that the selected covariates explain 23.72% of the total discrepancy for weekday schedules and only 7.34% for weekend schedules (Table 5.5[7]). Globally, there is a statistically significant effect of the selected variables on schedule discrepancy. Professional status, profession type, and age are the variables with the strongest effect on schedule dissimilarities for weekday schedules (explaining, respectively, 15.1%, 12.9%, and 10.2% of the total

6 Ibid.

7 Ibid.

discrepancy). Similarly, even though their explanatory power is rather low, these variables also have the strongest effect in explaining the discrepancy of weekend schedules (explaining, respectively, 2.7%, 2.2%, and 2.3% of the total discrepancy). In the same way, a multifactor discrepancy analysis was conducted with respect to the customized cost setting of substitutions operations (Table 5.4b[8]). The obtained results show similar effects of socio-demographic variables (Table 5.6[9]). Therefore, the different cost settings chosen for this study do not have an impact on dissimilarity analysis. The default setting will be maintained for the rest of the study.

As explained in section 5.3.4, to evaluate the schedule assignment method, travel-activity data are divided into "training" and "test" sets, representing, respectively, 877 and 363 individuals with their activity schedules. Based on the same selected variables, we displayed a regression tree analysis on the training data set. We identified eight clusters for weekday schedules (Figure 5.7a) and six clusters for weekend schedules (Figure 5.7b). For weekday schedules, the regression tree generated essentially three groups of schedules corresponding to working people (clusters 1–3), students (clusters 4 and 5), and the non-working people (clusters 6–8). Then, we can observe differences between working people since young adults (clusters 2 and 3) seem to start their working day earlier than the elders (cluster 1). Similarly, the principal variability between students' schedules depends on their age: scholars' schedules (cluster 4) clearly show two peaks of "school" activity, while older scholars (cluster 5 students older than 18) have less homogenous schedules with more "leisure" and "shopping" activities. Finally, gender has an effect on activity patterns of non-working people, including retired, unemployed, and house husband/wife. Men seem to perform more "leisure" activities (cluster 6) and women seem to perform more shopping activities (clusters 7 and 8).

For weekend schedules, the age is the most relevant variable splitting the schedules into two main groups: younger people schedules – from 18 to 28 years old – corresponding to the cluster 1, and a group of older people combined with children (less than 18 years old). The first group exhibited schedules with an important number of "leisure" and "visiting" activities showing one peak of activities at the beginning of the afternoon and continued late at night. However, the second group presents differences related to some professions within which people are susceptible to work

8 Ibid.

9 Ibid.

during the weekend such as shop or business owners, manual laborers, and farmers (clusters 2 and 3). Cluster 4 represents very young people with schedules containing essentially "leisure" activities. Finally, clusters 5 and 6 represent elderly who have no "working" activities during the weekend.

Based on regression tree results, we extracted (IF-THEN) rules to attribute schedules to individuals, depending on their socio-demographic variables. For every individual, in the test data set, we identify the cluster corresponding to his socio-demographic profile and we sample randomly one schedule from those belonging to the cluster. In order to improve the schedule sampling method, we identified the most central schedule in each cluster (called the medoid sequence), which represents the schedule with the lowest distance to the other schedules in the cluster [GAB 13]. Then we computed a distance between every schedule and the central schedule. This distance is then used as a weight parameter to supervise the schedule sampling in a way to have less probability to select schedules showing a large difference from the central schedule (non-typical schedules). Figures 5.8a and 5.8b represent an example of a pair of observed and assigned weekday and weekend schedules obtained from the IF-THEN rules extracted from the regression tree analysis.

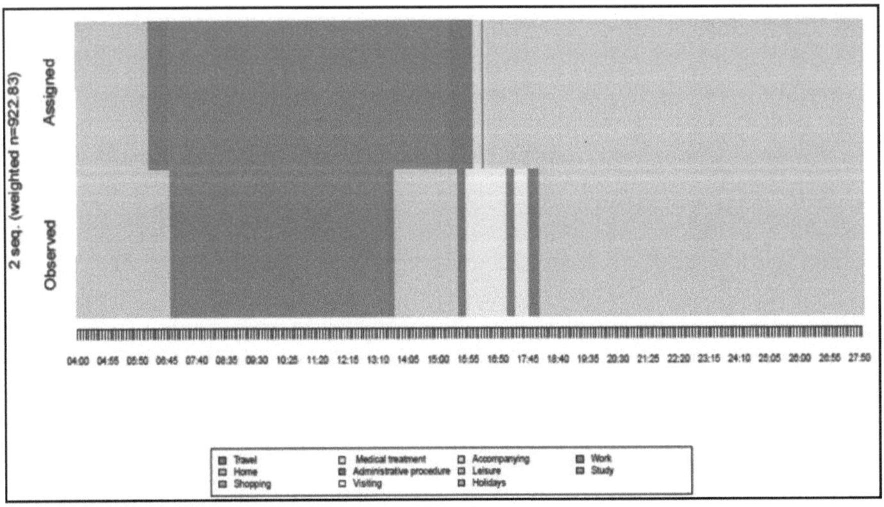

Figure 5.8a. *Example of a pair of observed and assigned weekday schedules*

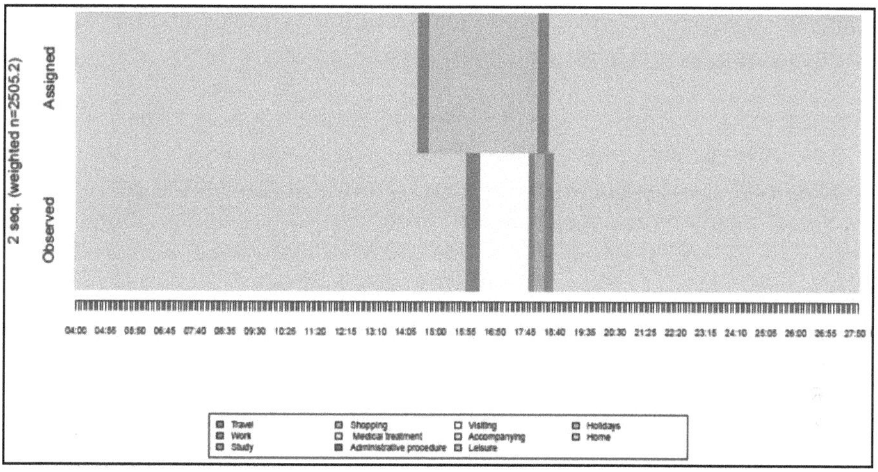

Figure 5.8b. *Example of a pair of observed and assigned weekend schedules*

In order to evaluate the performance of using the socio-demographic variables for assigning activity schedules, we computed, for every individual, in the test data, a dissimilarity distance between his "observed" schedule and the "assigned" schedule. The average dissimilarity distance between individual observed schedules and assigned schedules is of 80 for weekday schedules and 138 for weekend schedules (the maximal dissimilarity value is equal to 288 corresponding to the total number of elements/states by sequence and retrieved when two sequences have no similar element). Schedule assignment performance can be assessed by comparing the dissimilarity distance between observed and assigned schedules ($d_{Observed/Assigned}$) with the average distance between every observed schedule and all the other schedules ($d_{Observed/All\ schedules}$). Hence, the schedule assignment can be considered efficient when $d_{Observed/Assigned}$ is significantly lower than $d_{Observed/All\ schedules}$.

The results of the comparison indicate that week schedules assigned to individuals are globally similar to their observed schedules since the average $d_{Observed/Assigned}$ is less important than the average $d_{Observed/All\ schedules}$ ($d_{Observed/Assigned} = 80 < d_{Observed/All\ schedules} = 110$). However, weekend schedules attribution is less efficient ($d_{Observed/Assigned} = 138 > d_{Observed/All\ schedules} = 111$). This difference is mainly due to a lower effect of socio-demographic variables on schedule dissimilarity of weekend schedules. Therefore, it would be important to identify and include other variables that can enhance the explanation of weekend schedules' dissimilarities.

In addition, we remark that the schedule attribution performance depends on the socio-demographic classes. For weekend schedules, the third class (corresponding to young working people) exhibits the lowest performance since the median distance between observed and assigned schedules is $d_{Observed/Assigned} = 133$, while the median distance between the observed and other schedules is $d_{Observed/All\ schedules} = 118$ (Figure 5.9a).

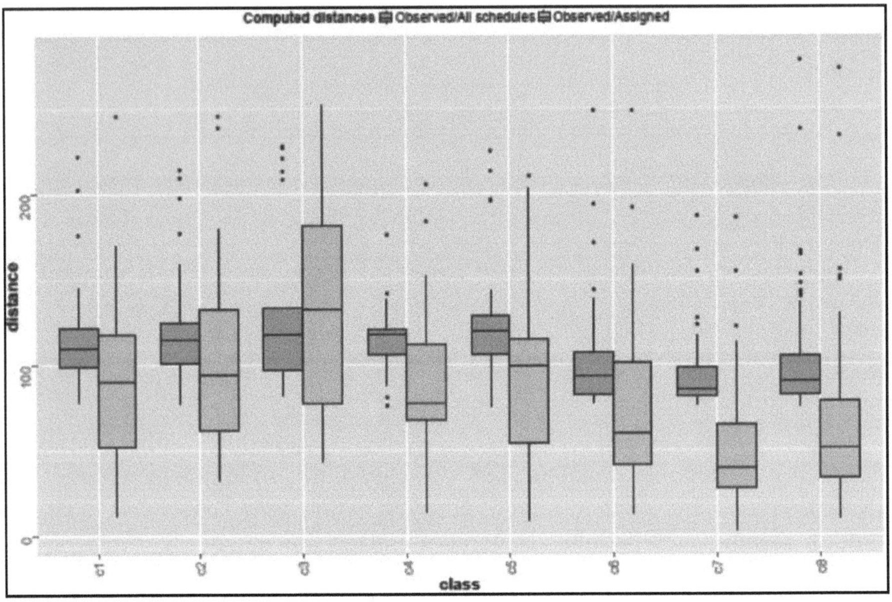

Figure 5.9a. *Distributions of the computed average distances $d_{Observed/Assigned}$ (orange color) and $d_{Observed/All\ schedules\ (Red\ color)}$ (blue color) in every class of weekday schedules. The boxplots represent the quartiles. The horizontal line represents the medians and the points represent extreme values. For a color version of this figure, see www.iste.co.uk/lutoff/hydrometeo2.zip*

This difference may be explained by the high variability in the within-class schedules. Concerning weekend schedules, the performance of the assignment is low for the different classes of schedules (Figure 5.9b).

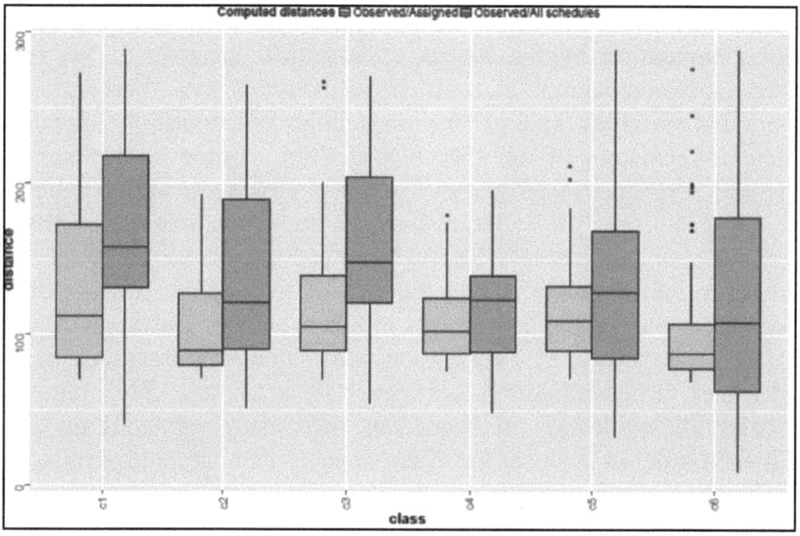

Figure 5.9b. *Distributions of the computed average distances* $d_{Observed/Assigned}$ *(orange color) and* $d_{Observed/All\ schedules\ (Red\ color)}$ *(blue color) in every class of weekend schedules. The boxplots represent the quartiles. The horizontal line represents the medians and the points represent extreme values. For a color version of this figure, see www.iste.co.uk/lutoff/hydrometeo2.zip*

5.5. Conclusion and discussion

The objective of this study is to provide schedule attribution rules based on the effect of socio-demographic characteristics on schedules dissimilarities. These rules are defined for improving an existing trip-based mobility model proposed by Debionne *et al.* [DEB 16] in order to assess motorists' exposure to road flooding in a French flood-prone region. Previous studies on the impact of extreme weather disruption on travel-activity behavior highlighted the importance of travel motivation on travel adaptation and activity rescheduling decisions. Therefore, an activity-based approach is needed for simulating travel-activity patterns and evaluating population exposure to road flooding during their mobility.

Activity-based approach has gained a growing attention in travel demand modeling since it takes into consideration travels' motivations by postulating that travel is a demand derived from individuals' needs to perform activities distributed in space. This approach implies a shift in focus from trips to activities and individual schedules, which requires the consideration of the sequential character of travel-activity behavior. Hence, in activity-based travel demand models, individual schedules are usually simulated for a synthetic population by the means of statistical analysis of the relation between socio-demographics and travel-activity data. Various socio-demographic variables (age, gender, household constraints, etc.) are pointed as relevant predictors of mobility behavior by looking at the number of travels and performed activities by category. However, some studies have explicitly examined the relationship between socio-demographics and activity sequencing process. Recent applications of Sequences Analysis methods for analyzing life course trajectories showed the benefits of those methods for exploring hidden patterns in life sequences. Furthermore, Studer *et al.* [STU 11] developed a regression tree method for assessing the effect of covariates on dissimilarities between state sequences. We proposed, in this study, an application of sequences discrepancy analysis on activity schedules in order to evaluate the effect of socio-demographic variables. The main objective of this study is the identification of IF-THEN rules enabling the attribution of weekday and weekend activity schedules to a population described by their socio-demographics for an implementation in a flood exposure model. For instance, working people having more than 45 years old can be linked to a schedule from cluster 1 in the regression tree of weekday schedules (Figure 5.10).

The analysis of individual schedules shows a difference between weekday and weekend travel behaviors regarding travels' numbers and daily activity participation. Furthermore, some socio-demographic variables (age, gender, and professional status) have an important effect on travels' behavior in term of trip frequencies and durations. The consideration of activity sequences leads to the identification of the temporal distribution of activity participation during the day. A discrepancy analysis is used for evaluating the effect of socio-demographic variables on activity sequences similarity. We identified the relevant variables displaying significant statistical effect on schedules discrepancy within two different activity substitutions costs parameterization. In our case, costs seem not to influence discrepancy results. The classification of activity sequences based on the effect of the selected variables generated eight clusters of weekday schedules where professional status and age seem to be the most discriminating variables. On the other hand, six clusters of weekend schedules were displayed corresponding essentially to different class of ages and profession types. The regression tree results were used for assigning a sampled schedule from each cluster to individuals based on their corresponding socio-demographic profile. The evaluation of schedule assignment model shows that weekday schedules assignment is more efficient than weekend schedules assignment due to the smaller effect of socio-demographic variables on weekend schedule discrepancy. Furthermore, the performance of the assignment rules depends on the socio-demographic profile, since some clusters exhibit larger differences between observed and assigned activity sequences. The heterogeneity level in schedules belonging to each cluster may be explained by the within clusters schedules variability.

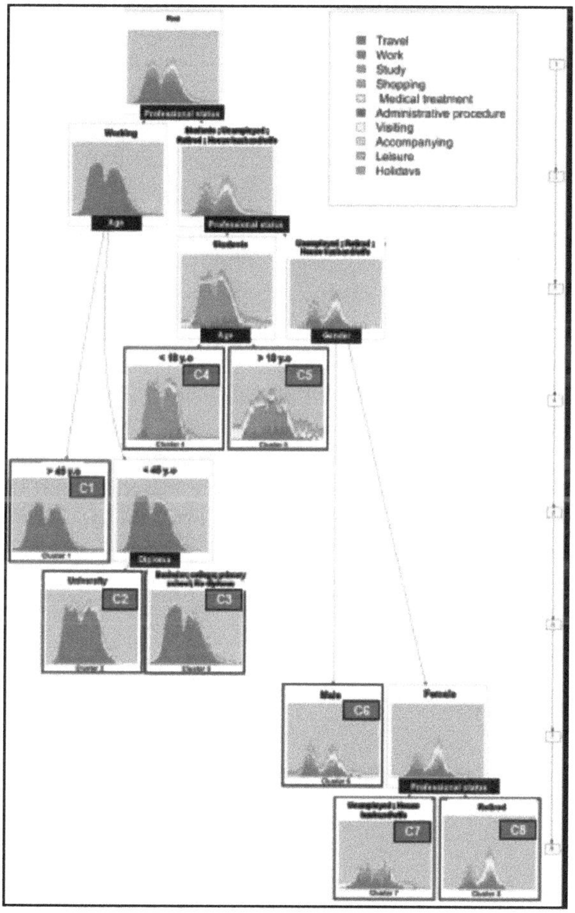

Figure 5.10. *Regression tree results for weekday schedules. The regression tree works as follows. All individual schedules are grouped in the first node. A discrepancy analysis is displayed to identify the variable explaining the greatest part of schedules discrepancy (with the highest* R^2*). The schedules are then partitioned based on this variable in such a way that the resulting child nodes are as much as possible homogeneous (with low within dissimilarity). This operation is repeated recursively until no significant effect of socio-demographic variables is registered in nodes' schedules*

Although the present application of sequence analysis methods on activity schedules provides new insights about the relationship between socio-demographic characteristics and travel-activity behavior, several improvements for future work remain. First, by considering only an average weekday schedule and an average weekend schedule, the variability of mobility behavior between different weekdays is not taken into account. Second, two costs setting parameters (based on activities' flexibility) were implemented in this study and showed no distinct effect in dissimilarity results. Third, different costs setting parameters regarding other dimensions of activity scheduling are necessary for the improvement of the Optimal Matching distance computation. Finally, travel-activity data were divided into "training" and "test" sets to enable the evaluation of the proposed schedule assignment method. This evaluation technique presents some limitations regarding data sets sampling and the reduction of data quantity. An improvement of the evaluation technique could be achieved by combining different sources of mobility data.

The findings of this chapter contribute to enhance knowledge about the interaction between socio-demographic characteristics of individuals and their travel-activity behavior. Methodologically, this study showed both benefits and limitations of using sequences analysis techniques for exploring activity schedules. The resulting schedule assignment rules, provided by the regression tree analysis, can then be employed for improving the MobRISK model, enabling the simulation of individual travel-activity patterns and the assessment of individual exposure to road flooding during a flash flood event in the Gard region.

5.6. References

[ABB 86] ABBOTT A., FORREST J., "Optimal matching methods for historical sequences", *The Journal of Interdisciplinary History*, vol. 16, no. 3, pp. 471–494, 1986.

[AND 01] ANDERSON M.J., "A new method for non-parametric multivariate analysis of variance", *Austral Ecology*, vol. 26, no. 1, pp. 32–46, 2001.

[AND 16] ANDRE-POYAUD I., CHARDONNEL S., CHARLEUX L. *et al.*, "Mobility & activity patterns of individuals and parenting couples in the metropolitan area of Grenoble", Miscellaneous, Halshs-01260244, 2016.

[ANT 01] ANTOINE J.-M., DESAILLY B., GAZELLE F., "Les crues meurtrières, du Roussillon aux Cévennes", *Annales de Géographie*, vol. 622, pp. 597–623, 2001.

[ARE 00] ARENTZE T., TIMMERMANS H., *Albatross: A Learning Based Transportation Oriented Simulation System*, Eirass, Eindhoven, 2000.

[ASH 08] ASHLEY S.T., ASHLEY W.S., "Flood fatalities in the United States", *Journal of Applied Meteorology and Climatology*, vol. 47, pp. 805–818, 2008.

[AUL 09] AULD J., MOHAMMADIAN A., "Framework for the development of the agent-based dynamic activity planning and travel scheduling (ADAPTS) model", *Transportation Letters*, vol. 1, no. 3, pp. 245–255, 2009.

[BAL 06] BALMER M., AXHAUSEN K., NAGEL K., "Agent-based demand-modeling framework for large-scale microsimulations", *Transportation Research Record: Journal of the Transportation Research Board*, vol. 1985, pp. 125–134, 2006.

[BAR 95] BARRETT C., BERKBIGLER K., SMITH L. *et al.*, "An operational description of TRANSIMS", LA-UR-95-2393, 1995.

[BAT 88] BATAGELJ V., "Generalized Ward and related clustering problems", in BOCK H.H. (ed.), *Classification and Related Methods of Data Analysis*, North-Holland, Amsterdam, pp. 67–74, 1988.

[BEN 98] BEN-AKIVA M.E., BOWMAN J.L., "Activity based travel demand model systems", in MARCOKE P. *et al.* (eds), *Equilibrium and Advanced Transportation Modelling*, Springer US, pp. 27–46, 1998.

[BHA 99a] BHAT C.R., KOPPELMAN F.S., "Activity-based modeling of travel demand", in HALL R.W. (ed.), *Handbook of transportation Science*, Springer US, pp. 35–61, 1999.

[BHA 99b] BHAT C.R., MISRA R., "Discretionary activity time allocation of individuals between in-home and out-of-home and between weekdays and weekends", *Transportation*, vol. 26, no. 2, pp.193–229, 1999.

[BHA 04] BHAT, C., GUO, J., SRINIVASAN, S. *et al.*, "Comprehensive econometric microsimulator for daily activity-travel patterns. Transportation Research Record", *Journal of the Transportation Research Board*, vol. 1894, pp. 57–66, 2004.

[BIL 06] BILLARI F.C., FÜRNKRANZ J., PRSKAWETZ A., "Timing, sequencing, and quantum of life course events: a machine learning approach", *European Journal of Population/Revue Européenne de Démographie*, vol. 22, no. 1, pp. 37–65, 2006.

[BLA 11] BLANCHARD P., FILLIEULE O., "Sequence analysis for political science", paper prepared for the 2011 Annual Meeting of the American Political Science Association, September 1–4, 2011.

[BOW 97] BOWMAN J., BEN-AKIVA M., Activity-based travel forecasting, Activity-Based Travel Forecasting Conference, Washington, DC, US Department of Transportation, Report DOT-97–17, 1997.

[BOW 01] BOWMAN J.L., BEN-AKIVA M.E., "Activity-based disaggregate travel demand model system with activity schedules", *Transportation Research Part A: Policy and Practice*, vol. 35, no. 1, pp. 1–28, 2001.

[CHA 74] CHAPIN F.S., *Human Activity Patterns in the City: Things People do in Time and in Space*, Wiley, New York, 1974.

[COO 10] COOLS M., MOONS E., CREEMERS L. *et al.*, "Changes in travel behavior in response to weather conditions: do type of weather and trip purpose matter?", *Transportation Research Record: Journal of the Transportation Research Board*, vol. 2157, pp. 22–28, 2010.

[DEB 16] DEBIONNE S., RUIN I., SHABOU S. *et al.*, "Assessment of commuters' daily exposure to flash flooding over the roads of the Gard region, France", *Journal of Hydrology*, vol. 541, pp. 636–648, 2016.

[ELI 08] ELIAS W., NEWMARK G., SHIFTAN Y., "Gender and travel behavior in two Arab communities in Israel", *Transportation Research Record: Journal of the Transportation Research Board*, vol. 2067, no. 2067, pp. 75–83, 2008.

[ETT 93] ETTEMA D., BORGERS A., TIMMERMANS H., "Simulation model of activity scheduling behavior", *Transportation Research Record*, vol. 1413, pp. 1–11, 1993.

[ETT 09] ETTEMA D., VAN DER LIPPE T., "Weekly rhythms in task and time allocation of households", *Transportation*, vol. 36, no. 2, pp. 113–129, 2009.

[FOS 16] FOSSET P., ANDRE-POYAUD I., BANOS A. *et al.*, "Exploring intra-urban accessibility and impacts of pollution policies with an agent-based simulation platform: GaMiroD", *Systems*, vol. 4, no. 1, p. 5, 2016.

[GAB 13] GABADINHO A., RITSCHARD, G., "Searching for typical life trajectories applied to childbirth histories", in LEVY R., WIDMER E. (eds), *Gendered Life Courses – Between Individualization and Standardization. A European Approach Applied to Switzerland*, LIT, Vienna, pp. 287–312, 2013.

[GAR 94] GÄRLING T., KWAN M.P., GOLLEDGE, R.G., "Computational-process modelling of household activity scheduling", *Transportation Research Part B: Methodological*, vol. 28, no. 5, pp. 355–364, 1994.

[GAR 98] GÄRLING T., KALEN T., ROMANUS J. *et al.*, "Computer simulation of household activity scheduling", *Environment and Planning A*, vol. 30, no. 4, pp. 665–679, 1998.

[GAU 09] GAUTHIER J.A., ERIC D.W., PHILIPP B. *et al.*, "How much does it cost? Optimization of costs in sequence analysis of social science data", *Sociological Methods and Research*, vol. 38, no. 1, pp. 197–231, 2009.

[GOL 94] GOLLEDGE R.G., KWAN M.P., GARLING T., *Computational-Process Modelling of Household Travel Decisions Using a Geographical Information System*, University of California Transportation Center, Berkeley, CA, 1994.

[HAG 70] HÄGERSTRAND T., "What about people in regional science?", *Papers in Regional Science*, vol. 24, no. 1, pp. 7–24, 1970.

[HAM 50] HAMMING R.W., "Erro-detecting and error-connecting codes", *Bell System Technical Journal*, vol. 29, pp. 147–160, 1950.

[HAN 81] HANSON S., HANSON P., "The travel-activity patterns of urban residents: dimensions and relationships to sociodemographic characteristics", *Economic Geography*, vol. 57, no. 4, pp. 332–347, 1981.

[JOH 01] JOH C.H., ARENTZE T.A., TIMMERMANS H.J., "Multidimensional sequence alignment methods for activity-travel pattern analysis: a comparison of dynamic programming and genetic algorithms", *Geographical Analysis*, vol. 33, no. 3, pp. 247–270, 2001.

[JOH 11] JOH C.H., TIMMERMANS H., "Applying sequence alignment methods to large activity-travel data sets: heuristic approach", *Transportation Research Record: Journal of the Transportation Research Board*, vol. 2231, pp. 10–17, 2011.

[KRU 83] KRUSKAL J.B., "An overview of sequence comparison: time warps, string edits, and macromolecules", *SIAM Review*, vol. 25, no. 2, pp. 201–237, 1983.

[KWA 14] KWAN M.P., NINGCHUAN X., GUOXIANG D., "Assessing activity pattern similarity with multidimensional sequence alignment based on a multiobjective optimization evolutionary algorithm", *Geographical Analysis*, vol. 46, pp. 297–320, 2014.

[LES 09] LESNARD L., KAN M.Y., Two-stage optimal matching analysis of workdays and workweeks, Sociology, Working Papers, Department of Sociology, University of Oxford, UK, 2009.

[LES 14] LESNARD L., "Optimal matching and social sciences", accessed December 22, available at: https://halshs.archives-ouvertes.fr/halshs-00008122/document, 2014.

[LEV 66] LEVENSHTEIN V.I., "Binary codes capable of correcting deletions, insertions, and reversals", *Soviet Physics Doklady*, vol. 10, pp. 707–710, 1966.

[LU 99] LU X., PAS E.I., "Socio-demographics, activity participation and travel behavior", *Transportation Research Part A: Policy and Practice*, vol. 33, no. 1, pp. 1–18, 1999.

[MAN 06] MANLY B.F., *Randomization, Bootstrap and Monte Carlo Methods in Biology*, vol. 70, CRC Press, Boca Raton, 2006.

[MCN 95] MCNALLY M.G., "An activity-based microsimulation model for travel demand forecasting", in ETTEMA D., TIMMERMANS H. (eds), *Activity-Based Approaches to Transportation Modeling*, Elsevier Science, pp. 37–54, 1995.

[NAU 13] NAULIN J.-P., PAYRASTRE O., GAUME E., "Spatially distributed flood forecasting in flash flood prone areas: application to road network supervision in southern France", *Journal of Hydrology*, vol. 486, pp. 88–99, 2013.

[PAS 84] PAS E.I., "The effect of selected sociodemographic characteristics on daily travel-activity behavior", *Environment and Planning A*, vol. 16, no. 5, pp. 571–581, 1984.

[PER 14] PERCHOU C., KESTENS Y., THOMAS F. *et al.*, "Assessing patterns of spatial behavior in health studies: their socio-demographic determinants and associations with transportation modes (the RECORD Cohort Study)", *Social Science & Medicine*, vol. 119, pp. 64–73, 2014.

[RAS 14] RASOULI S., TIMMERMANS H., "Activity-based models of travel demand: promises, progress and prospects", *International Journal of Urban Sciences*, vol. 18, no. 1, pp. 31–60, 2014.

[REC 86] RECKER W.W., MCNALLY M.G., ROOT, G.S., "A model of complex travel behavior: part I—theoretical development", *Transportation Research Part A: General*, vol. 20, no. 4, pp. 307–318, 1986.

[RUI 14] RUIN I., LUTOFF C., BOUDEVILLAIN B. *et al.*, "Social and hydrological responses to extreme precipitations: an interdisciplinary strategy for postflood investigation", *Weather, Climate and Society*, vol. 6, no. 1, pp. 135–153, 2014.

[SMI 95] SMITH L.R., BECKMAN D., ANSON K. *et al.*, TRANSIMS: TRansportation ANalysis and SIMulation System, Los Alamos National Laboratory Unclassified Report, LA-UR-95-1664, Los Alamos, NM, 1995.

[STU 11] STUDER, M., RITSCHARD, G., GABADINHO, A. *et al.*, "Discrepancy analysis of state sequences", *Sociological Methods and Research*, vol. 40, no. 3, pp. 471–510, 2011.

[SHA 12] SHARIF H.O., HOSSAIN M.M., JACKSON T. *et al.*, "Person-place-time analysis of vehicle fatalities caused by flash floods in Texas", *Geomatics, Natural Hazards and Risk*, vol. 3, pp. 311–323, 2012.

[TER 16] TERTI G., RUIN I., ANQUETIN S., GOURLEY J.J., "A situation-based analysis of flash flood fatalities in the United States", *Bulletin of the American Meteorological Society*, vol. 98, no. 2, pp. 333–345, 2016.

[TIM 02] TIMMERMANS H., ARENTZE T., JOH C.H., "Analysing space-time behaviour: new approaches to old problems", *Progress in Human Geography*, vol. 26, no. 2, pp. 175–190, 2002.

[VAU 99] VAUGHN K.M., SPECKMAN P., SUN D., Identifying relevant socio-demographics for distinguishing household activity-travel patterns: a multivariate regression tree approach, Paper, National Institute of Statistical Science, Washington, DC, 1999.

[VER 10] VERSINI P.-A., GAUME E., ANDRIEU H., "Assessment of the susceptibility of roads to flooding based on geographical information–test in a flash flood prone area (the Gard region, France)", *Natural Hazards and Earth System Science*, vol. 10, no. 4, pp. 793–803, 2010.

[WIL 98] WILSON W.C., "Activity pattern analysis by means of sequence-alignment methods", *Environment and Planning A*, vol. 30, pp. 1017–1038, 1998.

6

Geolocated Tweets as a Means of Observing Extreme Natural Events. First Specifications[1]

6.1. Introduction

In recent years, research has shown a growing interest in information from geolocated digital footprints created by users of social networks, particularly on the Twitter microblogging platform. This form of social media allows users to create and exchange information in short messages, sometimes accompanied by photographs, videos, or other media. For the researcher, Twitter offers two advantages: the principle of communication is based on the diffusion of short and concise messages; in addition, the company Twitter is freely made available to any Internet user, by opening an account on the platform and a few lines of code, some of the content broadcast daily, via different APIs[2]. A decade of scientific work has developed a wide spectrum of fields of interest and disciplines involved [STE 15]: the first publications relating to the use of Twitter appeared the year after the platform was launched and focused on the observation of the

Chapter written by Camille CAVALIÈRE, Paule-Annick DAVOINE, Céline LUTOFF and Isabelle RUIN.

1 The authors would like to thank the members of the SLIDE team at the Laboratoire d'Informatique de Grenoble for the implementation of the tweets collection server.

The work presented in this chapter is part of the ANR MobiClimEx project (ANR-12-SENV-0002-01) and the MISTRAL-HyMeX project.

2 An API (Application Programming Interface) is a library of functions that provides services to other applications or software.

microblog as a *social phenomenon* [JAV 07]: it was then a question of understanding the construction of user networks, and the reasons that engage them in a process of using and adopting these tools in their daily lives.

At the end of the 2000s, the gradual democratization of devices integrating a geolocation system (GPS) prompted Twitter to give users connected from their smartphones the opportunity to add geographical coordinates to their tweets [CEB 16]. This initiative marked a new era for research: the researcher gained the ability to collect information created in real time by human sensors deployed in space [MAC 12], [SCH 13], and the metadata[3] accompanying the tweet are proving to be invaluable indicators for researchers who focus their attention on the triptych *what, when, where* [DEL 09].

The work undertaken on the study of geolocated tweets has thus highlighted the ability of tweets to reveal the spatio-temporal uses of urban spaces [AND 13], the mobility of individuals on an international scale [HAW 14], or the spatio-temporal characteristics of an unusual event occurring in the real world and recorded on the network [LUC 16]. It must be said that the scientific community makes this observation repeatedly: because of its particular properties, a geolocated tweet offers the possibility of taking a new look at the phenomena for which the human and social sciences had little (if any) data to quantify and qualify them [ELW 12]. With regard to natural hazards, the work carried out has demonstrated the relevance of these particular data as a source of spatio-temporal information for the diagnosis and characterization of crisis periods [DEL 09]. While it can now be taken for granted that a geolocated tweet informs us about the individual's centers of interest and reflects his spatial practices [AND 13], it is nevertheless an *imperfect* and *biased* digital footprint. It is imperfect because the majority of tweets sent by the users are collected and redistributed without their knowledge [QUE 16]. It is also imperfect because the data we download through the API is delivered outside any specific geographical and social *context*, specific to the time of its production [QUE 16]. It is biased because tweeting is above all an activity linked to a certain category of the population that lives and travels within certain particular spaces, these varying according to hours and days [CEB 16], [LI 13]. In this chapter, we try to highlight the different pitfalls that the researcher will face when work is based on geolocated digital footprints. The methodology we

3 According to the current formula, metadata is data that describes the data: they provide information on the quality, conditions of use, update dates, etc., of the downloaded datasets.

are developing is based on the complementarity between data sets of various natures and abductive reasoning[4]. It is applied to extreme hydro-meteorological events – known as the *cévenols* episodes – that affected eight departments in Southeastern France (Drôme, Ardèche, Lozère, Vaucluse, Gard, Var, Bouches-du-Rhône, and Hérault) between October 2014 and January 2015. These flash floods were characterized by intense convective storms that caused short episodes of torrential rainfall resulting in runoff from watersheds and sudden rises in river levels. The rest of the chapter is structured as follows: the first two parts present the use of geolocated tweets in the theme of natural hazards and contextualize the issues related to the use of geolocated digital footprints; the third part presents the data used and the methodology adopted for the construction of the crisis tweets. The results of the case study are presented in the last part of the chapter.

6.2. Geolocated tweets: a geographical opportunity or form of risk-taking?

While the research was particularly prolific at the beginning of 2010, enthusiasm for geolocated digital footprints seems to be drying up recently: and indeed, the literature review published by Steiger *et al.* [STE 15] indicated that the themes of study and methods implemented were having difficulty in renewing themselves.

6.2.1. *How have geolocation devices changed the individual's place in real-time risk management?*

Over the past decade, the massive adoption by populations of social media platforms and new communication tools integrating geolocation devices has changed the phase of real-time crisis management [MEI 15]. The citizen-as-sensor [GOO 07] then became the key witness to real-world events, insofar as the simple fact of having a device equipped with an on-board GPS makes it possible to record and locate immediately, on a virtual network, any unusual or dangerous phenomenon. Consequently, any individual involved and connected can produce relevant and valuable field information in the same way that a group of individuals witnessing a phenomenon is able to produce a data set "more accurately than an expert"

4 Abductive reasoning differs from the hypothetical-deductive approach traditionally used in science in which it involves inferring hypotheses from the observed facts.

[GOO 10], who is usually sent to the field after the phenomenon has occurred [DAS 14]. In the area of natural hazards and disasters, the approach based on early detection of phenomena and real-time crisis management has remained a major focus of the work undertaken.

Real-time crisis management is based on an *operational approach* based on a top-down/bottom-up dual logic [MAC 12]: individuals who witness a phenomenon disseminate localized information by telephone calls, SMS, or tweets containing a recipient statement. This field information is collected and cross-checked with physical and social data to verify its reliability and to organize assistance to vulnerable populations [MEI 15]. All the information compiled is fed back into an interactive crisis map [MAC 12]. The Peta Bencana project is an example of the implementation of this dual approach: it arose directly from the need for the municipal authorities of the Indonesian capital city Jakarta of more than 30 million inhabitants, frequently subjected to flooding, to have information acquired in real time on the ground, in order to detect and locate flooded areas and then to organize assistance for the populations. The original project is now extended to three other megalopolis of the Java Island (Bandung, Semarang and Surabaya). Peta Bencana is an information collection platform created through crowdsourcing and tweet applications. The latter are central to the project: since 2015, all Southeast Asian countries have experienced strong growth in terms of social network users; moreover, the average time spent daily on social networks in Indonesia was estimated at 03:16 per person in 2017[5] (the same year, it was estimated at 01:23 for France). Users are given precise instructions (use geographical hashtags[6] to spatialize information, estimate water levels) in order to create information that can be used directly by the authorities. For each flood occurrence, these data are mapped and cross-checked with topographic, meteorological, demographic, and socio-economic data sets to verify the validity of information from the participating public and to study the impact of phenomena in different areas and on different population categories [MEI 15].

5 https://wearesocial.com/fr/blog/2017/01/digital-social-mobile-les-chiffres-2017 (last accessed on September 06, 2017).

6 A hashtag is a keyword or a set of keywords preceded by the sign #: it identifies a tweet to a specific theme (natural phenomenon, cultural, sporting, political event, etc.).

6.2.2. *Useful information or Big Garbage[7]?*

This work is not part of this operational approach, but rather of the questions surrounding the effective benefits of geolocated tweet material for the study of a geographical and social problem. Certainly, geolocated tweets have already proven their usefulness in the issue of natural hazards: research carried out in post-crisis situations has demonstrated their ability to provide information that evolves according to the stage of the phenomenon – before the crisis, during, and after [BLA 14], [HER 14]. They also demonstrate that their spatial and semantic components correspond to the location and real effects of the phenomenon [DAS 14]. But can we dig deeper?

The availability of geolocated tweets has indeed aroused such enthusiasm that it inspires the feeling, upon reading the publications produced in early 2010, that these footprints would revolutionize the world of research and mark the advent of a new paradigm of knowledge construction [AND 08], [KIT 13], [MIL 15]. All for good reason: digital footprints, including geolocated tweets, were quickly integrated into the Big Data phenomenon, the term researchers use to designate the masses of footprints left more or less consciously by Internet users during browsing or captured by the intelligent devices that accompany individuals on a daily basis, and whose properties exceed the technical capabilities of our traditional storage and analysis methods. Although the definition and scope of Big Data are still unclear [GOO 13], [KIT 13], the authors agree to define their main characteristics according to the 3V formula: volume, velocity, and variety. The daily flow of tweets sent out on a global scale meets these three criteria: volume, as 500 million tweets created on a daily basis require unprecedented storage capacities, acquisition speed, facilitated by the very nature of the tweet and the spontaneity of users, and the diversity (variety) of content. A tweet is indeed a form of soft data that refers to all information created and disseminated on the Internet, and not controlled by state agencies, as opposed to hard data, which can be assimilated to traditional institutional data whose acquisition procedures are controlled and documented [AUD 14].

7 Term introduced by Rakesh Agrawal to designate the masses of data created daily without structure or constraint, and whose analysis must be preceded by a significant cleaning phase: https://icdm2016.eurecat.org/wpcontent/uploads/2016/05/WaveFour-ICDM-Dec-2016-Rakesh.pdf (last accessed on September 06, 2017).

Although geography, and more particularly the theme of natural hazards, only deals with a tiny fraction of these Big Data [DAS 14], [HER 14], [SAR 15], it does not escape a series of pitfalls that have appeared sporadically in the literature in recent years: Goodchild warns that there is a lack of information about the representativeness of these new data and the constraints that their availability imposes on the spatial questions posed by geographers [GOO 13]. Audard *et al.* [AUD 14] question the reliability and certainty of experimental results from data over which researchers have no control, since they are absent from the data acquisition and publication process. Steiger *et al.* [STE 15] attest, among other things, to the lack of methods that cross-reference tweets and other data sources. Finally, Quesnot [QUE 16] argues that the strong potential accorded to digital footprints would ultimately only prove to be smoke and mirrors, these footprints being – because of their conditions of acquisition and availability – unable to capture the meaning of the relationships maintained by the individual with the space traveled.

Are digital footprints then of no interest to the researcher? Steiger *et al.* [STE 15] make an irrefutable observation: tweets are useful because of their high spatial and temporal resolution and can be a support material for the acquisition of reliable knowledge in various fields of application.

6.3. Properties of geolocated tweets

In practice, geolocated tweets are part of the footprints created by the daily use of a Web service. At a time when digital technology is taking root in society, it seems essential to us to understand the characteristics of these footprints and the challenges related to their analysis.

6.3.1. *Who tweets when activating geolocation and why?*

However, any geolocated tweet is above all a trace of social bias. Indeed, active participation in social networks depends, in the first place, on adherence and motivation, either two factors that can be closely linked to the individual's age, origin and social conditions, accessibility to digital networks (both financial and technical), level of education, etc., or all variables whose effects have repercussions on overall tweeting activity and activity in response to a phenomenon. But what does the intention to activate geolocation reveal[8]?

We have only a few suggestions to highlight the demographic and social characteristics of the population that activates this feature on Twitter. Li *et al.* [LI 13] attempted to establish social profiles of the categories of these particular users in California, based on a combination of geolocated tweets, demographic and socio-economic data. The authors found a strong positive correlation between the level of education, the field of professional activity and salary, on the one hand, and the high concentration of tweets, on the other hand (the hotspot on Twitter in California actually corresponds to Silicon Valley, the birthplace of the Web giants). However, there is no guarantee to the researcher that the results observed in the Californian population will not apply to other regions of the same country or the world: unlike survey data, a set of tweets is not a representative sample of a parent population [GOO 13]. More recently, Cebeillac and Rault sought to identify populations that create geolocated content in a specific location, in a luxury shopping mall in India, frequented by the upper-middle class [CEB 16]. In this context, it appeared that the creation of geolocated content was more closely associated with the average income of this class, i.e. with individuals who frequented the place to achieve social promotion through a luxurious environment but who did not necessarily have the financial means to consume in this place. On the other hand, the real customers of the shopping mall differed from this ostentatious logic and were not very active on social networks.

Therefore, while we can assume that geolocated tweets reflect users' concerns in a specific location, we do not know if these concerns are shared by all people who frequent the location in question.

6.3.2. *Digital footprints with heterogeneous forms*

The geolocated tweet is presented as a protean digital footprints, which, in addition to the spatio-temporal metadata associated with it, provides qualitative information. In our field of study in southeastern France, first

8 Adding a location to a tweet is a voluntary act, since the user must first allow Twitter to add such metadata in his account settings, and then must again allow the application to enable the smartphone's GPS for each tweet broadcast.

analysis and categorization of the tweets collected identified three main forms of geolocated tweets emitted in the context of flash floods:

– The *official footprint* – the tweet was issued from the account of a civil protection actor, a public actor, or a weather forecasting organization: *#Nîmes: services de la ville placés en crise inondation, cadereaux en crue et ruissellement urbain abondant.#alertemétéo @tvsud* (#Nîmes: city services placed in flood crisis situation, flooded *cadereaux* (torrents) and abundant urban runoff. #weatheralert @tvsud) issued via the account of a TVSud journalist;

– The thematic footprint – the tweet evoked behaviors, concerns, instructions, etc., *Alerte orange ! Cours annulé, on rentre à la maison!* (Amber alert! Class canceled, we're going home!), Il fait nuit, il pleut, c'est alerte orange mais je fais les magasins (It's dark, it's raining, it's amber alert but I'm shopping).

– The *observation footprint of the physical phenomenon:* the tweet translated the observed environmental conditions into real time. The observation was objective "It is raining"; "There are beginning to be storms," or reflected the user's personal feeling about the intensity of the observed phenomenon: *Nonnnn comment il déluge* (Noooo the rain is torrential), *Olala je suis seule et y a grave de l'orage je vais mourir là...??*(Oh dear I am alone and there is a severe storm, will I die here...??).

As the few examples added to illustrate this classification have shown, geolocated tweets in response to a damaging natural phenomenon contain tweets from official actors, revealing an intention to disseminate clear and collective information on the virtual network. The *Peta Bencana* project mentioned above is based on this operating principle: the individual who disseminates information to the competent authorities is aware of the reuse of the same information. However, not every geolocated tweet issued in times of crisis has the same intention: the other two categories of tweets identified in our case are more like contextual information, sometimes intended for talking about oneself, just like every day tweets do. In most cases, the user does not intend to disseminate rich and accurate information for reuse for scientific purposes. This is why Harvey [HAR 13] describes as contributed data (rather than Volunteered Geographic Information) all spatial or geolocated information created by Internet users as part of their personal use of social media platforms, information that is not necessarily created to meet a need from a research community (or other). There is then a dichotomy between the user's will and the scientist's rigorous spatial need.

Nevertheless, tweets centered on the individual is not without interest: among the above-mentioned tweets, one of them reveals an individual behavior that is probably dangerous (see "It's dark, it's raining, it's amber alert but I'm shopping"), while others, depending on the vocabulary chosen or the tone given, give us details on the intensity felt by their authors in the face of the phenomenon (see "Noooo the rain is torrential" or "Oh dear I am alone and there is a severe storm, will I die here...??").

6.3.3. *Quantities of tweeted digital footprints*

To this qualitative and social heterogeneity, we must add the quantitative biases that introduce additional constraints into the themes and places of analysis. First, Twitter provides free access to only a small percentage of the total traffic captured by the platform: between 1% and 3% of flows [QUE 16], [SLO 15]. The tweets returned to the client by the API are, therefore, subject to sampling whose conditions remain ambiguous[9] [QUE 16], although Twitter guarantees the non-sensitivity of its algorithm to the metadata associated with the tweets and, therefore, a balanced representation of all types of tweets [SLO 15]. In a second step, the tweets streamed with GPS geolocation represent only a minimal proportion of the global flow siphoned daily on the network: Brovelli *et al.* [BRO 14] estimate that the proportion of tweets transmitted with longitude/latitude coordinates varies between 1% and 2% of the total daily transmitted tweet stream. On the other hand, it is difficult to obtain more precise or more recent figures: for Hawelka *et al.*, it is rather 1% of the flow [HAW 14], whereas for Sloan and Morgan, this proportion decreases to 0.85% [SLO 15]. It would, therefore, be risky to say the least to claim that a set of tweets captures the totality and complexity of the processes that take place in a space (which is also in interaction with other spaces). Tweeting is a one-time phenomenon and not an act performed in a regular time interval; similarly, tweets do not reach all users of a space since not all communicate via social media. It seems essential to us to highlight here this paradox that seems specific to geosciences: the abundance of Big Data is in no way synonymous with exhaustiveness [KIT 13], [SLO 15].

In addition, the amount of footprints produced in response to an event is also related to multiple social factors: where the physical sensor network

9 However, if the customer submits a request with very selective criteria (precise keywords, tweets geolocated by GPS only), he may escape this sampling constraint.

records a quantitative and regular measurement (such as a seismic sensor network), a group of users traveling through a space on a daily basis form a mobile network on the same space; the density of this network varies according to the territories, regardless of the spatial scale considered, hours and days [AND 13]. The reviewed studies show that the spatial and temporal distribution of tweets is irregular, and depends on the spatial distribution of the population, in space and according to time [LUC 16]. Thus, phenomena that occur in sparsely populated areas may be under-represented or virtually absent from the network: the less populated the area, the less likely it is to encounter active users (not to mention the fact that peripheral or sparsely populated areas certainly do not have the same quality of connectivity as in urban areas). Similarly, at a local scale, population densities vary with time: if a phenomenon occurs in the middle of the day in a residential area, during a period of low occupancy, it is highly likely that it will not be recorded on the network.

The problem of geolocated digital footprints created in response to a natural phenomenon adds new obstacles: phenomena, whose physical characteristics are different, do not cause identical reactivity on the part of Twitter users, and, consequently, do not have the same diffusion impact on the network. Some major phenomena have a very high media impact and have the capacity to generate considerable tweet flows (hurricane, earthquake, etc.). On the contrary, local-scale phenomena do not lead to such a mobilization of users on the network.

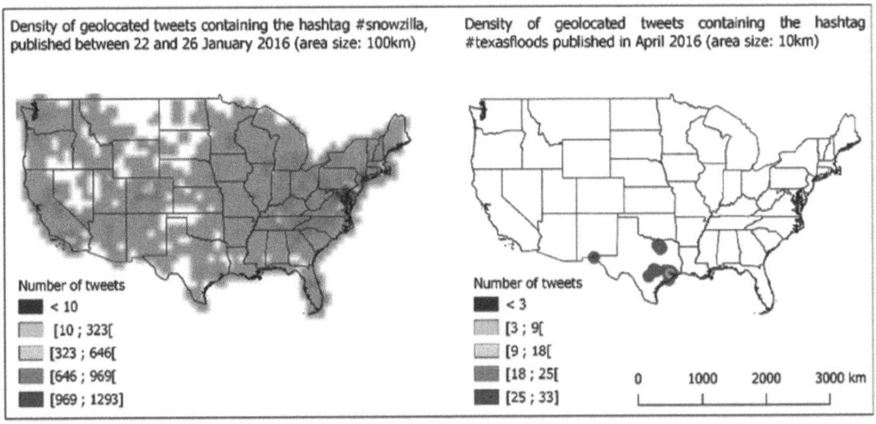

Figure 6.1. *Quantity of geolocated crisis tweets and spatial envelope of the diffusion of hashtags according to the type of phenomenon. For a color version of this figure, see www.iste.co.uk/lutoff/hydrometeo2.zip*

This is shown in Figure 6.1, which results from tests we conducted on two events in the United States. The map on the left shows the distribution of geolocated tweets mentioning the hashtag "#Snowzilla" in January 2016 (in direct reference to the Jonas blizzard). This phenomenon was widely publicized worldwide: in the United States, the hashtag spread far beyond the space physically affected by the storm. The map on the right shows the spatial distribution of the hashtag "#TexasFloods" in reference to the April 2016 flood events in Texas: here, the virtual envelope of the phenomenon did not exceed the envelope of the affected physical space.

6.4. Construction of a set of crisis tweets and mobilized data

This new section describes the different steps in the construction of the crisis tweets for flash floods; it also introduces additional data to provide elements of comparison from an official source.

6.4.1. *Methodology applied*

We built the crisis tweets set from a raw tweets set, respecting two constraints: that it was as complete as possible, in terms of tweet size, and that it was as accurate as possible in terms of information quality, and minimizing residual noise[10] in the final crisis tweets set.

6.4.1.1. *Collecting tweets*

The raw data set was extracted from a server connected to the Twitter Streaming API[11], which collects, cleans, and stores geolocated tweets from around the world. When collecting tweets in real time, retweets, i.e. tweets created by one user and re-broadcast by others on the network, are deleted to avoid data redundancy. The tweets are finally stored in a PostgreSQL database[12].

10 Residual noise refers to tweets that may contain vocabulary apparently related to flash flood phenomena, but used in a completely different context.

11 https://dev.twitter.com/streaming/overview.

12 We must point out that the server accidentally disconnected from the API on October 12, 2014 and that the collection of tweets was voluntarily interrupted from December 13, 2014 to January 4, 2015 (due to the lack of maintenance staff).

The extraction by a bounding box[13] covering the entire territory studied returned a raw data set consisting of 806,954 geolocated tweets issued in our eight departments between October 8, 2014 (the date on which the server was put into service) and January 31, 2015. Each tweet was characterized by the following attributes: a tweet ID number, user ID number, tweet content, creation date and time, and GPS coordinates. This dataset was the raw tweets set.

6.4.1.2. *Extraction of geolocated crisis tweets*

Potentially useful tweets can be selected according to their content [HER 14]: we compiled a glossary of weather-related keywords. Some words were considered necessary and were widely adopted by users, such as the glossary related to weather disturbances. The keywords were, therefore, not chosen at random but required a brainstorming step, during which we tried to inventory as much vocabulary (or lexical associations) as possible relating to the event.

Second, we needed to imagine the consequences and scenarios that could be envisaged during such events: electricity power cuts, road closures, isolated houses, the intervention of the fire brigade, uprooted trees, etc. The objective was to select tweets related to crisis periods, but which would not explicitly mention weather elements (such as thunderstorms or floods). By proceeding in this way, we expected tweets such as *j'ai plus d'électricité chez moi* (I have no electricity at home) and *j'ai plus d'électricité chez moi #orage #intempéries* (I have no electricity at home #storm #badweather) to be stored in the final crisis data set. Our research collected a set of 80 keywords (including lexical associations such as water + rising) divided into different categories (Figure 6.2). It should also be noted that in addition to the word rain, we were also looking for the verb in the present *pleut* (raining) (only the conjugation in the present tense was retained in order to promote the event at the time the tweet was created).

In a second step, we queried the data in an unsupervised approach: we used text mining tools to cluster[14] the words used in the tweets. This made it possible to identify a new vocabulary directly related to the hydrometeorological event that we had not considered or whose spelling was incorrect (Figure 6.3).

13 A bounding box is a geometric shape that delimits the influence of a space of interest.

14 Lexical clustering highlights words that appear frequently associated in tweets.

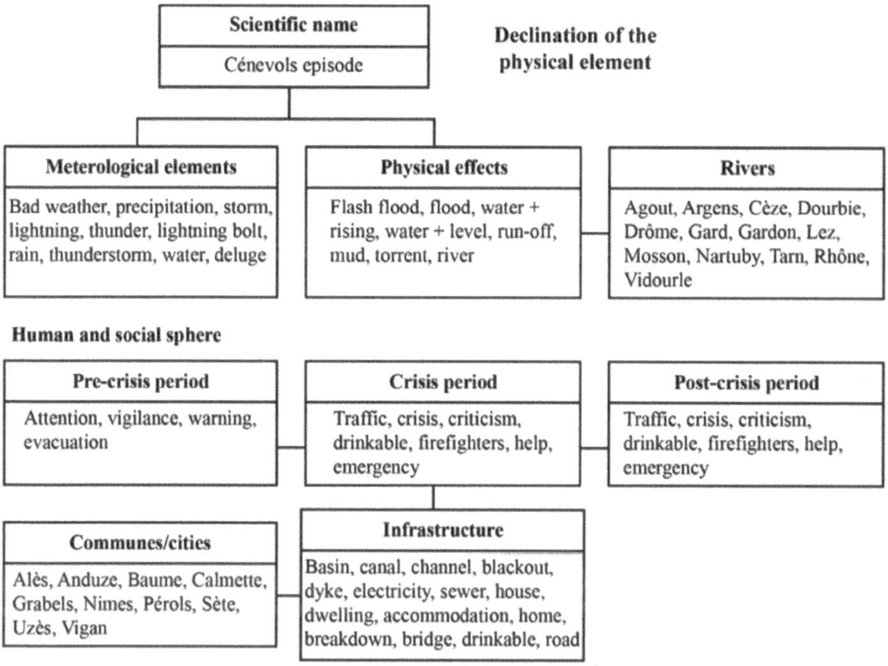

Figure 6.2. *Inventory of keywords searched with translations provided below each box*

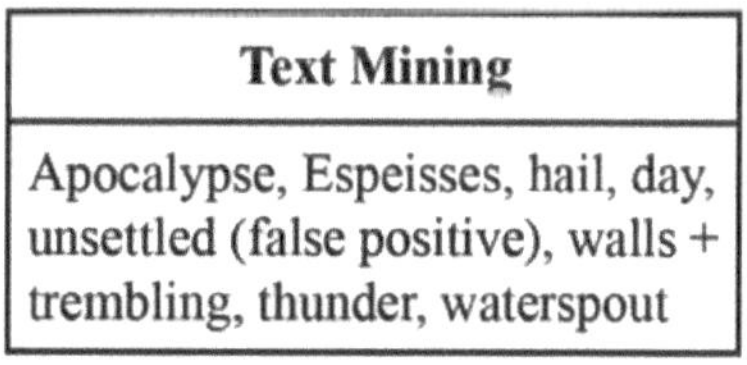

Figure 6.3. *Keywords highlighted by text mining*

A certain number of the keywords chosen or discovered by text mining could be found in various contexts: *pompiers, secours, alerte, and inundation* (firefighters, help, warning, and flood); we can mention firefighters or help in case of a road accident, inadvertently flooding your bathroom, and so on. We, therefore, considered it essential to check the correspondence between the extracted tweets and the event context. However, we did not have algorithms reliable enough to handle small datasets. This data review and selection step was therefore carried out manually. Since many tweets can be subject to interpretation, we defined a set of rules for validating tweets that were considered relevant. All these conditions are indexed in Table 6.1:

Purpose of the tweet	Subject of the tweet	Example of a tweet
Phenomenon	Weather bulletin or alert	*#Nîmes: nouvel épisode de pluies violentes à l'instant. Crues et débordements observés ou envisagés dans de nombreux secteurs. #alertemétéo* (#Nîmes: another episode of heavy rain just now. Floods and rising water observed or forecast in many areas. #weatheralert)
	Current or occurring event	*Demain j'ai pas cours, mon lycée a été inondé !!* (Tomorrow I don't have class, my high school got flooded!!!)
	Consequences (people, property, infrastructure)	*Le pont qui mène en ville est plein d'eau, j'suis coupé du monde !??* (The bridge that leads into town is flooded, I'm cut off from the world!)
Individuals faced with a phenomenon	Anxiety/fear	*Il pleut de dingue chez moi omg j'ai peur* (It's raining crazy heavy where I live omg I'm scared)
	Behavior	*Il fait nuit, il pleut, c'est l'alerte orange mais je fais les magasins* (It's dark, it's raining, there's an amber alert but I'm shopping)
	Opinion (crisis management, relief)	*@meteofrance merci pour cette #alerterouge qui n'a servi à rien. Pas une goutte de pluie* (@meteofrance thank you for this #redalert which was useless. Not a drop of rain)

Table 6.1. *List of conditions and examples of the selected tweets*

These conditions were defined through reading the tweets: we first classified them according to the purpose of the tweet (individual or phenomenon). The *Phenomenon* category refers to objective observations of phenomena in progress or occurring; the *Individuals* category contains information that is sometimes subjective and focused on the sole author of the tweet. It should be noted that some tweets can be difficult to assign a specific category: "The bridge that leads into town is flooded, I'm cut off from the world!" This tweet contains both an observation of the surrounding environment and a subjective feeling.

6.4.2. *Introduction of additional data: dates and severity of events*

Since 2011, the CatNat[15] database has been recording data on various natural disasters – landslides, storms, floods, forest fires – around the world. We consulted data to obtain the precise dates of the cévenols episodes that occurred in the fall of 2014 in our study area and noted the occurrence of six major events with their severity index provided with the data, evaluated from 1 to 5, an index of five corresponding to the most violent events (Table 6.2).

Start date	End date	Severity index
October 09, 2014	October 13, 2014	3
November 03, 2014	November 04, 2014	2
November 09, 2014	November 11, 2014	2
November 14, 2014	November 15, 2014	3
November 24, 2014	November 27, 2014	2
November 27, 2014	November 30, 2014	3

Table 6.2. *Temporalities and severity of events in Southeastern France, fall 2014*

We obtained three sets of data for analysis: raw tweets and crisis tweets allowed us to see how tweeting activity was structured in our field; CatNat data provided us with a basis for comparison between virtual digital footprints and official data.

6.5. First explorations and analyses of data sets

This section presents the results of the methodology for building the crisis tweets set, presented above, as well as the first spatial, temporal, and thematic analyses carried out.

15 http://www.catnat.net/.

6.5.1. *Characteristics of the geolocated crisis tweets sets*

The next two paragraphs present the characteristics of the crisis tweets set extracted from the defined method.

6.5.1.1. *The low virtual impact of bad weather*

The first undeniable observation was that the geolocated tweets linked to the occurrence of a damaging hydrometeorological phenomenon were only a handful involved in the virtual routine: the crisis tweets set represented only 0.38% of the raw tweets set extracted from the field of study and throughout the period considered (from October 2014 to January 2015). It should also be noted that some keywords did not return any results (mainly the names of rivers or cities). Tweets mentioning the elements of weather disturbances formed the majority: for the crisis set, tweets containing weather information (keywords in the "weather elements" box in Figure 6.2: *pluie, pleut, déluge, précipitations*, orage, éclair, tonnerre (rain, raining, deluge, precipitation, storm, lightning, thunder, etc.) represented 77.3% of the information extracted. Both datasets are mapped in Figure 6.4.

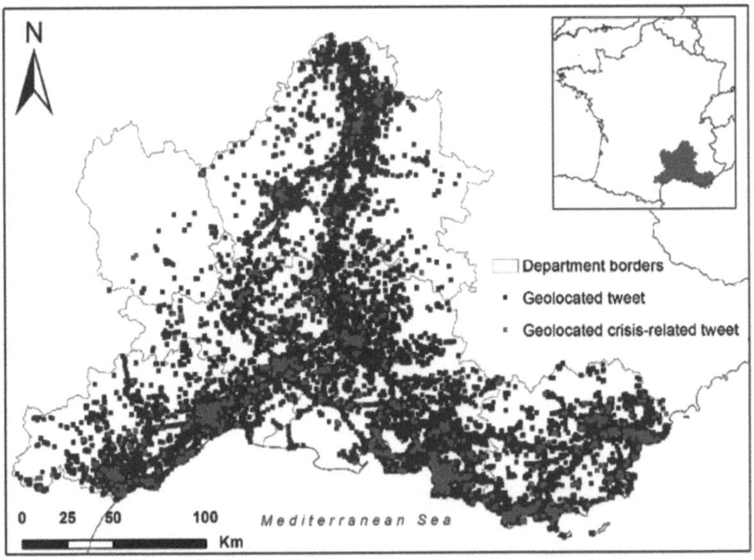

Figure 6.4. *Mapping of the raw (blue) and crisis (red) tweet sets. For a color version of this figure, see www.iste.co.uk/lutoff/hydrometeo2.zip*

The location of tweets, whether raw or crisis in nature, reflected the location of populations: we could immediately identify the Rhône corridor as well as the network of Mediterranean coastal cities. Thus, the territories whose footprint remained the strongest in virtual space corresponded to urban environments. As for rural areas, they could remain empty of geolocated tweets (this was the case in particular in the Vercors drômois, the north of Lozère and the Camargue) or presented some occasional clusters of geolocated tweets (Ardèche, Drôme provençale).

6.5.1.2. *User tweeting practices*

The raw tweets set has 18,563 users; 1,330 users, or 7.16%, which helped to create information related to hydrometeorological crises. The involvement of these users in the production of crisis information, therefore, remains marginal. The quantitative use of Twitter seems homogeneous since the majority of users only tweeted a few times in four months. In the raw tweets set, 50% of users wrote less than five tweets and 80% less than 25. In the crisis tweets set, 57% of users sent only one tweet and 90% less than five tweets during the five months of collection. In addition to these average profiles, we identified a minority of users, who sent, on average, nearly 500 tweets during the period in question and who represent 1.7% of the users of the raw data set (10,218 is the maximum number of tweets for a single user). In the crisis tweets set, only 1.34% of users sent more than 15 tweets.

6.5.1.3. *Temporalities of daily tweeting activity*

For both sets of data (raw and crisis), we compared the daily temporal distribution of tweets for the entire period collected. Since the daily numbers varied greatly from one set of tweets to another, we converted these numbers into percentages: for each day recorded, this percentage corresponds to the ratio between the number on the day and the total number on the corresponding month (Figure 6.5).

The daily variability of the recorded flows was low on the raw tweets set, while the crisis set seemed particularly sensitive to the occurrence of weather events. In fact, the observable crisis tweets peaks coincided with three periods reported by CatNat data (indicated by the black arrows in Figure 6.5): October 10 and 11, November 4, and November 25–28. Smaller peaks could be distinguished for other crisis dates, between November 9 and 11 and November 14–15. There was also a time lag between the official start of the crisis and the response time of Twitter users: for example, November 24 marked the beginning of a phenomenon inventoried in the CatNat database but only recorded a handful of crisis tweets (on that day, the phenomenon was perhaps very localized, and/or outside an urban area?). The temporal distribution of crisis tweets had other peaks that did not correspond to any event reported by CatNat: early December 16 and 19, 2014 and January 21, 2015.

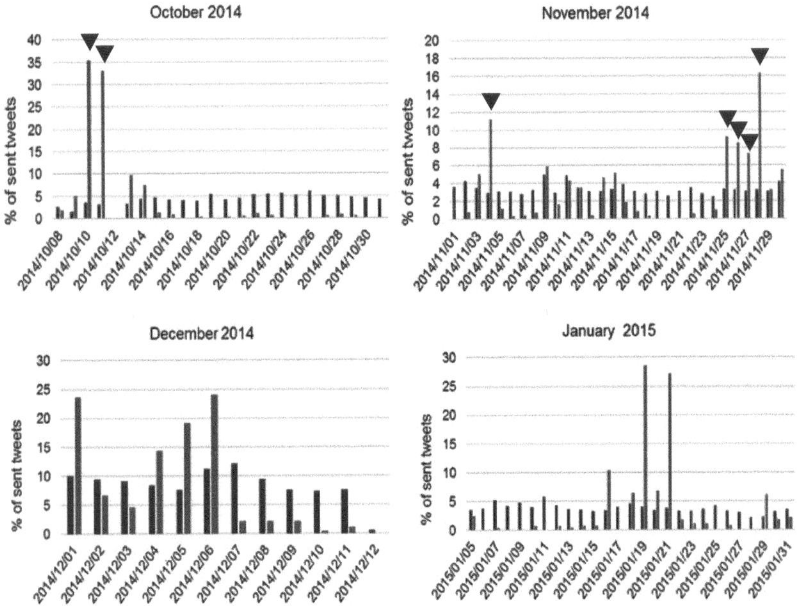

Figure 6.5. *Daily tweets flow per month for raw tweets (in blue) and crisis (in red) sets. Black triangles indicate the occurrence of an extreme hydrometeorological event. For a color version of this figure, see www.iste.co.uk/lutoff/hydrometeo2.zip*

There was also a persistent background activity related to the weather: crisis tweets were regularly sent out without any report of major phenomena, which probably means that Twitter users are sensitive to weather disturbances. On the other hand, we cannot affirm the existence of a correlation between the severity of the crises and the corresponding tweet flows: for example, the CatNat data indicate that the severity of the crises that took place on November 14 and 15, then from 27 to 30 November was assessed at the same level (3/5) but the latest crisis generated higher tweets flows. The recurrence of events a few days apart can potentially explain this increase in tweets for the second event.

We have thus verified the detection and signal potential of tweets with respect to hydrometeorological disturbances in our field of study. The next issue was to determine what the high spatial concentrations of geolocated crisis tweets reveal.

6.5.2. *Spatio-temporal and semantic analysis by abduction*

The following results present the spatial–temporal and semantic analyses performed from the raw tweets and crisis sets; it presents an example of an abduction-inferred hypothesis.

6.5.2.1. *Detection of activity areas*

First, we analyzed the spatial and temporal distribution of tweets by studying standardized densities: the objective being to identify potential activity hotspots by erasing the effects of population densities, based on the study of the set of raw tweets. We were thus seeking to detect tweeting anomalies, i.e. spaces that concentrate more tweets than inhabitants, potential markers of a significant event within this particular space.

We used Kernel density[16] to generate monthly rasters of tweet densities from the raw set. A population density raster was created from INSEE's squared data[17] from the 2010 census. The square data divide the metropolitan territory into squares of 200 m in length and record the number of inhabitants associated with each unit. The final rasters were generated by calculating the ratio between the monthly tweets density rasters and the population density raster, and represented by a collection of maps. This represented the spatial distribution of the ratios calculated for each of the months collected and highlighted the areas where densities of tweets were higher than population densities (Figure 6.6).

The areas where the pixel values were between 0 and 1 corresponded to spaces where the activity could be qualified as standard, i.e. the densities of tweets were less than or equal to the population density. Areas where this

16 From a seedling of points such as geolocated tweets, the Kernel density (or kernel method) creates a smoothed surface, in the form of a raster file (a matrix of pixels whose value corresponds to the probability estimate established from a search radius and a smoothing parameter).

17 http://www.insee.fr/fr/themes/detail.asp?reg_id=0&ref_id=data-carroyees.

ratio was greater than 1 formed our areas of interest and designated areas that had more concentrated tweets than inhabitants.

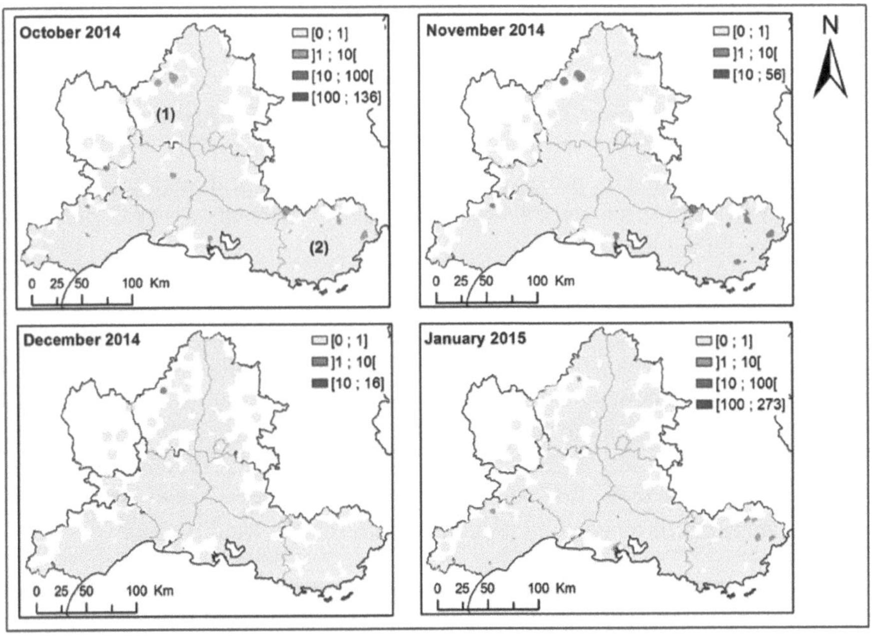

Figure 6.6. *Detection of monthly hotspots within the field between October 2014 and January 2015. For a color version of this figure, see www.iste.co.uk/lutoff/hydrometeo2.zip*

Such a representation makes it possible to detect hotspots that persisted in time and space. These hotspots were mainly located in the departments of Ardèche and Var, but not found in the main urban areas. It should be noted that cities generally remained the preferred places for digital expression and the use of social networks [LI 13]. The analysis of hotspots, therefore, made it possible to identify "anomalies" in flows in specific sectors or periods. To explain the existence of these anomalies, we inferred that these hotspots could correspond to the occurrence of unusual events, potentially of meteorological origin (in such spaces, it is probably less likely to encounter abnormal activity related to social or cultural events). We, therefore, proposed to explore the location of crisis tweet hotspots in order to visualize the possible existence of a correlation between the emission hotspots we detected and high densities of crisis tweets.

6.5.2.2. *Monthly densities of crisis tweets*

To do this, we explored the spatial and temporal variations in the monthly densities of crisis tweets (Figure 6.7). Not surprisingly, the highest densities of tweets were found in regional urban areas and along the Mediterranean coast, which have the highest population densities. We also saw the emergence of certain sources of crisis tweets that were repetitive over time, particularly in the northern Ardèche (1), the western Gard (3), and the Var (2).

By comparing these hotspots with those we identified in the previous step, we discovered common structures. We chose to explore in more detail the hotspot located in the north of the Ardèche, in order to test the relationship between these hotspots and the location of hydrometeorological events. In other words, we checked whether tweeting anomalies made it possible to identify places particularly affected by crises, outside urban areas.

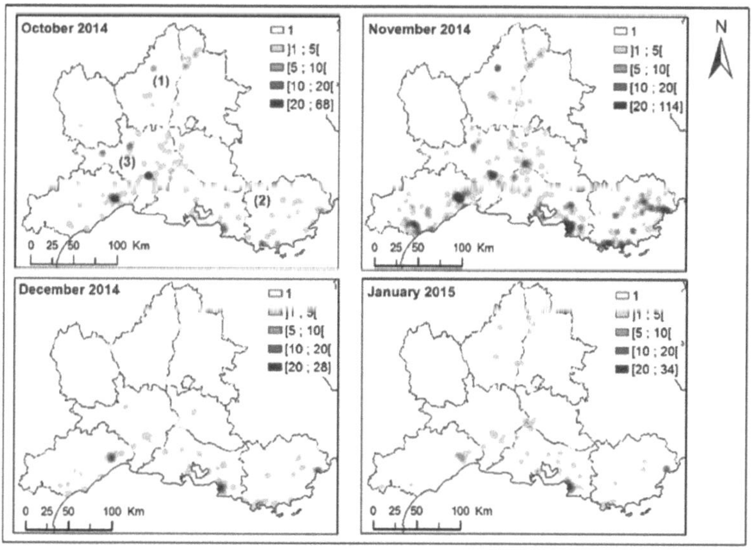

Figure 6.7. *Monthly densities of crisis tweets in the field between October 2014 and January 2015. For a color version of this figure, see www.iste.co.uk/lutoff/hydrometeo2.zip*

6.5.2.3. *Exploration of the Ardèche hotspot*

We wanted to know precisely what happened in this particular place over the time periods corresponding to the tweeting hotspot. By exploring this motive in more detail, it seemed to us to focus on a rural commune, Le Cheylard.

In this municipality, the period studied recorded 4,063 tweets sent by four users. Three of them were occasional users while the last one could be qualified as a super tweeter. He has written 3,972 tweets and the average daily number of tweets he sent was 43. The hotspot highlighted was, therefore, partly due to the super activity of a single user. We will not consider him as a spammer: this individual was indeed the user for whom we saw the maximum participation in the creation of crisis tweets. To explore this user's habits and behavior, we selected and analyzed the information contained in his 43 crisis tweets in order to classify them by theme. We selected four of them, summarized in Table 6.3 (the topics of the tweets are shown in bold).

Purpose of the tweet	Subject of the tweet	Example of a tweet
Local user environment	**Weather**: real-time observation	*Pluie et gris ici* (Rainy and gray here)
	Flooding: the user's location has been flooded	*Là, ça y est c'est l'inondation* (This is it, this is the flood)
	Alert: the Ardèche is on alert or weather watch	*noir ici, vigilance orange* (dark here, amber alert)
External object	**Information**: tweet is linked to a crisis but does not concern the local tweeter's environment	*Nouvel accident dans le Gard: un homme retrouvé noyé http://t.co/oBy8tM2X2C via BFMTV* (New accident in the Gard: a man found drowned http://t.co/oBy8tM2X2C via BFMTV)

Table 6.3. *Thematic classification and examples of crisis tweets concerning the Cheylard user (Ardèche, October 2014 to January 2015)*

Table 6.3 and Figure 6.8 (influence of each identified subject in the set of crisis tweets issued each day) indicate that this user's habits were identical to those we had previously identified: most of his crisis tweets were created between October and November, particularly during the crisis days identified in the CatNat data. In addition, we can learn that it was flooded on November 14 and 15. Much of the information created was semantically homogeneous, and therefore, provided little useful and relevant field information: when tweeting to describe a local event, this user created information related to the weather, and more specifically to an ongoing downpour. Nevertheless, it illustrates the ability of the individual to safeguard any phenomenon that occurs in the immediate environment, which escapes the experts. In addition, it should be noted that the object we have named "information" in Table 6.3, which represents a significant part of the total information created, refers to events that affected other territories, for example other departments or other cities mentioned in tweets, while geolocation indicates that the individual remained in his area of residence. We thus saw another problem, referred to as asynchronous telepresence [MIL 07], which qualifies the effects of advances in communication technologies regarding the emancipation of the individual's physical presence, in time and space, to become aware of a phenomenon. The Internet user no longer needs to be in the right place at the right time to follow, in real time or not, the progress of an event.

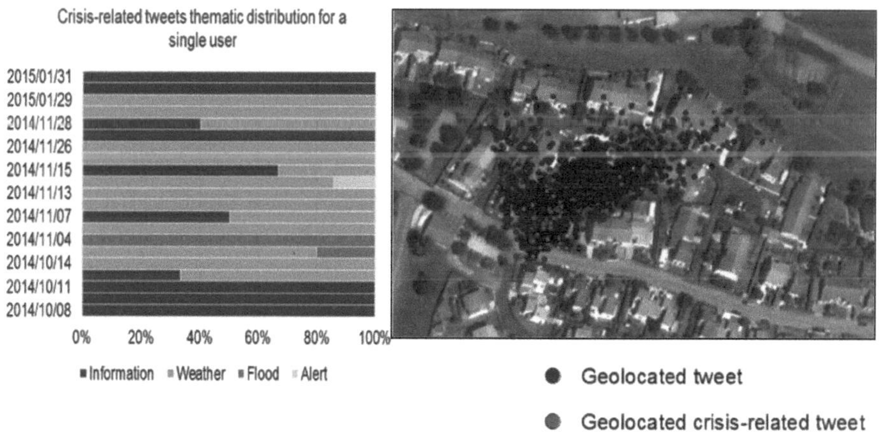

Figure 6.8. *Exploring the Cheylard user's tweets. For a color version of this figure, see www.iste.co.uk/lutoff/hydrometeo2.zip*

An abnormal activity was therefore not necessarily a reflection of an intense crisis; moreover, the local information sent by the identified user was not quantitatively sufficient to reconstruct the meteorological event that concerned him. This example reinforced the question of the representativeness of Tweet data: if a single individual who is virtually "hyperactive" provides the information to document a phenomenon in a specific place, how can its relevance for scientific analysis be verified? Indeed, an individual creating digital content on the Web from elements of the real world can select objects, situations, or places based on any criteria other than those we are trying to measure.

6.6. Conclusion

This chapter, therefore, promises to be a warning against the systematic use of geolocated digital footprints in geographical problems, and also against the idea that they can replace the geographer/geomatician's traditional data sets. We verified the alleged existence of tweeting anomalies, called hotspots, and found, contrary to what we had assumed, that they were on the margins of urban areas or in rural areas; after which we hypothesized that these observations could be the result of disturbances, supposedly of meteorological origin. The representation of crisis tweet densities tended to confirm this hypothesis since the areas affected by the events corresponded to areas of high crisis tweets emission densities. However, during the final test, it appeared to us that these hotspots could also correspond to the abundant activity of isolated users, which can be considered as aberrant. This led us to raise the following question: what is a major event for a Twitter user, that is, an event that is, individually or collectively, important enough to be discussed? Nevertheless, the spatio-temporal analysis of geolocated crisis tweets can reveal the rhythms of social responses to flash flood phenomena; it then requires certain preliminary precautions, particularly with regard to places and users that present a statistically aberrant profile. The work would also require exploring other areas of activity, in order to determine the possible repetitiveness of the Ardèche structure analyzed here.

We also identified another problem: while we had hypothesized that the geolocated tweet emitted by smartphones was focused on the user's local and immediate environment, we detected a user emitting information about a much larger geographical area, and about events that occurred in several other departments. In the end, after the implementation of these first processes, a series of other questions emerged: how to deal with crisis information not related to the user's present territory at the time of tweeting? What are the factors responsible for the variation in tweeter mobilization? Or why are there time lags between the dates and times of the phenomena and the reactivity on the network?

At this stage, with the limited data collected, it is not possible to confirm the possibility of reconstructing, in an exhaustive way, the complexity of a crisis from geolocated tweets. Moreover, in the face of these full tweets and tweets about empty spaces, and in the face of these individuals with an aberrant profile, it now seems essential to us that research nuances the ability of geolocated tweets to serve as a material for the construction of objective knowledge.

6.7. References

[AND 08] ANDERSON C., "The end of theory: the data deluge makes the scientific method obsolete", available at: http://www.wired.com/2008/06/pb-theory/, 2008.

[AND 13] ANDRIENKO G., ANDRIENKO N., BOSCH H. *et al.*, "Thematic patterns in georeferenced tweets through space-time visual analytics", *Computing in Science and Engineering,* vol. 15, no. 3, pp. 72–82, 2013.

[AUD 14] AUDARD F., CARPENTIER S., OLIVEAU S., "Les "big data" sont-elles l'avenir de la géographie [théorique et quantitative]?", *20e Biennale de géographie d'Avignon*, *Géopoint 2014*, pp. 1–4, 2014.

[BLA 14] BLANFORD J., BERNHARDT J., SAVELYEV A. *et al.*, "Tweeting and tornadoes", *11th International ISCRAM Conference*, University Park, USA, 2014.

[BRO 14] BROVELLI M.A., ZAMBONI G., MUÑOZ C.A. *et al.*, "Exploring twitter georeferenced data related to flood events: an initial approach", *AGILE'2014 International Conference on Geographic Information Science*, Castellon, Spain, 2014.

[CEB 16] CEBEILLAC A., RAULT Y.-M., "Contribution of geotagged Twitter data in the study of a social group's activity space", *Netcom*, vol. 30, no. 3/4, pp. 231–248, 2016.

[DAS 14] DASHTI S., PALEN L., HERIS M.P. *et al.*, "Supporting disaster reconnaissance with social media data: a design-oriented case study of the 2013 Colorado floods", *11th International ISCRAM Conference*, University Park, USA, 2014.

[DEL 09] DE LONGEVILLE B., SMITH R.S., LURASCHI G., "'OMG, from here, I can see the flames!': a case use of mining location based social networks to acquire spatio-temporal data on forest fires", *International Workshop on Location Based Social Networks*, Seattle, USA, 2009.

[ELW 12] ELWOOD S., GOODCHILD M., SUI D., "Researching volunteered geographic information: spatial data, geographic research and new social practice", *Annals of the Association of American Geographers*, vol. 102, no. 3, pp. 571–590, 2012.

[GOO 07] GOODCHILD M.F., "Citizens-as-sensors: the world of volunteered information", *GeoJournal*, vol. 69, no. 4, pp. 211–221, 2007.

[GOO 10] GOODCHILD M.F., GLENNON J., "Crowdsourcing geographic information for disaster response: a research frontier", *International Journal of Digital Earth*, vol. 3, no. 3, pp. 231–241, 2010.

[GOO 13] GOODCHILD M.F., "The quality of big (geo)data", *Dialogues in Human Geography*, vol. 3, no. 3, pp. 280–284, 2013.

[HAR 13] HARVEY F., "To volunteer or to contribute locational information? Towards truth in labeling for crowdsourced geographic information", in SUI S., ELWOOD S., GOODCHILD M.F. (eds), *Crowdsourcing Geographic Knowledge*, Springer, Dordrecht, The Netherlands, pp. 31–42, 2013.

[HAW 14] HAWELKA B., SITKO I., BEINAT E. *et al.*, "Geo-located Twitter as proxy for global mobility patterns", *Cartography and Geographic Information Science*, vol. 41, no. 3, pp. 260–271, 2014.

[HER 14] HERFORT B., PORTO DE ALBUQUERQUE J., SCHELHORN S.-J. *et al.*, "Exploring the geographical relations between social media and flood phenomena to improve situational awareness", in HUERTA J., SCHADE S., GRANELL C. (eds), *Connecting a Digital Europe through Location and Place*, Springer International Publishing, Heidelberg, Germany, pp. 55–71, 2014.

[JAV 07] JAVA A., SONG X., "Why we Twitter: understanding microblogging usage and communities", *Proceedings of the Joint 9th WEBKDD and 1st SNA-KDD Workshop*, pp. 56–65, 2007.

[KIT 13] KITCHIN R., "Big data and human geography. Opportunities, challenges and risks", *Dialogues in Human Geography*, vol. 3, no. 3, pp. 262–267, 2013.

[LI 13] LI L., GOODCHILD M.F., XU B., "Spatial, temporal and socioeconomic patterns in the use of Twitter and Flickr", *Cartography and Geographic Information Science*, vol. 40, no. 2, pp. 61–77, 2013.

[LUC 16] LUCCHINI F., ELISSALDE B., GRASSOT L. *et al.*, "Paris tweets, données numériques géolocalisées et évènements urbains", *Netcom*, vol. 30, no. 3/4, pp. 207–230, 2016.

[MAC 12] MC DOUGALL K., "An assessment of the contribution of volunteered geographic information during recent natural disasters. Spatially enabling government, industry and citizens: research and development perspectives", *Needham*, GSDI Association Press, Needham, USA, pp. 201–214, 2012.

[MEI 15] MEIER P., *Digital Humanitarians: How Big Data Is Changing the Face of Humanitarian Response*, CRC Press, Inc., Boca Raton, USA, 2015.

[MIL 07] MILLER H.J., "Placed-based versus people-based geographic information science", *Geography Compass*, vol. 10, no. 3, pp. 503–535, 2007.

[MIL 15] MILLER H.J., GOODCHILD M.F., "Data-driven geography", *Geojournal*, vol. 80, no. 4, pp. 449–461, 2015.

[QUE 16] QUESNOT T., "L'involution géographique: des données géosociales aux algorithmes", *Netcom*, vol 30, no. 3/4, pp. 281–304, 2016.

[SAR 15] SARAVANOU A., VALKANAS G., GUNOPULOS D. *et al.*, "Twitter floods when it rains: a case study of the UK floods in early 2014", *24th International Conference on World Wide Web*, Florence, Italy, May 2015.

[SCH 13] SCHADE S., DIAZ L., OSTERMANN F. *et al.*, "Citizen-based sensing of crisis event: sensor web enablement for volunteered geographic information", *Applied Geomatics*, vol. 5, no. 1, pp. 3–18, 2013.

[SLO 15] SLOAN L., MORGAN J., "Who Tweets with their location? Understanding the relationship between demographic characteristics and the use of geoservices and geotagging on Twitter", *PLoS ONE*, vol. 10, no. 11, p. e0142209, available at: https://doi.org/10.1371/journal.pone.0142209, 2015.

[STE 15] STEIGER E., PORTO DE ALBUQUERQUE J., ZIPF A., "An advanced systematic literature review on spatiotemporal analyses of twitter data", *Transactions in GIS*, vol. 19, no. 6, pp. 809–834, 2015.

7

Adaptation Paces – Physical Cursors for Action Analysis

7.1. Introduction

Generally, studying how individuals or groups adapt to the hazardous conditions created by the flooding of a river requires comparing the dynamics of flooding with the dynamics of action. The adaptive nature of an action can only be understood if the conditions to which the individual or group in question is exposed at the time of its action are precisely described – are they confronted with a danger that appears quickly or slowly, what are the precursor signs that indicate the imminence of danger, and what could be called post-cursor signs, marking the return to normality?

The term cursor refers to a "small moving part sliding along a ruler or compass used to make various adjustments" (http://www.cnrtl.fr). In the case of a flood, "small moving parts" are more or less fine changes in the environment: darkening of the sky, intensification of rain, for example, for pre-cursors, recession of water levels, and reopening of cut roads for post-cursors. What interests us here is comparing physical cursors with the modalities or adaptive capacities of a particular social group. This comparison is an interdisciplinary matter in the sense that we must outline the physical reality in a way that is useful for the interpretation of human action – what degree of environmental urgency presides over the action in each situation, and what information is provided by the environment to position the cursor?

Chapter written by Céline LUTOFF, Brice BOUDEVILLAIN, Jean-Dominique CREUTIN and Sonia CHARDONNEL.

Our proposal in this chapter is to explore how certain physical cursors can help improve understanding of the social processes of responding to an event. We, therefore, begin by discussing the identification of physical cursors that could be useful at the time of the crisis; then we propose to approach the inter-crisis and to question the possible mobilization of physical cursors in a longer-term perspective.

7.2. Why are physical cursors needed to study adaptation during a crisis?

In this first section, we propose to question the relevance of physical cursors to better understand the social processes of adaptation to the flood event. How can certain elements of the physics of the phenomenon be mobilized to better understand social responses?

7.2.1. *Flooding is a "villainous risk"*

Unlike earthquakes or forest fires, floods are a "familiar" threat, and as such, they make it difficult to identify the danger. As such, François Dedieu speaks of a villainous risk[1] [DED 07], [DED 15]. Indeed, regions subject to flash floods usually receive heavy rains that cause rivers to react in a "normally" surprising way. These are often intermittent rivers, with very low flows for most of the year, which are only sporadically indicated by a sudden rise in water levels, most of the time with no major consequences. It should be noted that these "normally" harmless floods are always accompanied by warnings from the meteorological services, which cover a large region, target the occurrence of severe thunderstorms in a fairly safe manner, but are unsuitable for the scales that will locally transform the thunderstorm into a catastrophic flood. It is then difficult to "unmask the risk" (see Chapter 3 of Volume 1 on flash flood forecasting).

The consequence of these difficulties to be anticipated at fine scales is that exposed populations must rely on their own resources to establish an overview of the situation and its possible evolution. Early warning signals then play a major role in the ability of individuals and communities to take the necessary actions to protect themselves. However, these signals must

1 For F. Dedieu, "villainous risk" is defined as a threatening phenomenon hidden under normal appearances. This proposal allows us to fall between the theories of "impairment" – we knew but we did nothing, and "inevitability" – we could not know.

also be clear and easily interpretable, which is not always the case, particularly because of their "familiar" nature.

7.2.2. *The multiple appearances of a flash flood*

It is quite easy to depict flooding of a slow-moving river such as the Seine or the Somme, to use examples from our national flood adaptation policy (see Chapter 5 of Volume 1). The sequence of meteorological (heavy rains) and hydrological (rising waters) pre-cursors that predict the danger is described in days or even weeks. In January 1910, for example, the Seine in Paris (44,000 km^2) reached its flood peak level in seven days, after 13 days of sustained rainfall [MAR 97]. This sequence resulted in a general and relatively synchronous increase in the hazard over the entire river system.

The appearance of a flash flood is quite different, it is multiple. Generating thunderstorms are known to have extreme rainfall intensity and heterogeneity. The intensity of the rains activates a very rapid response of the rivers in terms of runoff and, therefore, imposes a local urgency for action based on this rapidity. The heterogeneity of the rain causes a considerable disparity in the individual responses of the sub-basins, which will divide the area into multiple "flood scenes." A flood scene can be defined as a place where the urgency for human action is homogeneously marked by the local watercourse – the same time of flood initiation and the same rhythm of water rise. With the same thunderstorm, this emergency to action covers up to two orders of magnitude (i.e. from 1 to 100) between "flood scenes" sometimes a few hundred meters apart. Thus, during feedback on the flood that hit the Var department in 2011, an employee from the tourist office of Brignoles, ready to close her office when seeing the waters of the neighboring river rise (water rising phase of the main river), said she had been dissuaded by her manager based higher in the city, who noted that the water that had invaded her street had disappeared (water level recession phase of a minor affluent)[2].

This multiple appearances thus introduce a contradictory message into the social network from one place to another and rapidly changing from one moment to the next. It is easy to imagine the difficulty for the individual to

2 Elements from the interdisciplinary feedback [RUI 14] organized following the flood on November 6, 2011.

adapt the already difficult synchronization of all his "ordinary" activities to a constraint whose pace and spatiality are complex[3].

This difficulty is amplified in a "connected" world where social rhythms are based on a rapidly growing set of interactions that is seen as a difficulty in itself or even a source of anxiety [MEL 97]. Nevertheless, the introduction of new technologies must be taken into account, with the explosion of social networks constituting an additional agent in the organization of the social response to extreme natural phenomena [DOU 17], [LIN 11]. However, we will not discuss these aspects in this chapter[4].

7.2.3. *Physical cursors for action analysis*

Analyzing the adaptation of action to floods requires as precise a description as possible of the dangerous conditions prevailing at the time and place of the action. Our proposal is, first, to represent the change in the flood during the crisis on a single axis. This single axis makes it possible to stagger the different stages of the flood by going from the meteorological and hydrological pre-cursors of the crisis to the "post-cursors" that mark the return to a normal situation. The notion of normality is used here as opposed to the exceptional nature of the crisis situation. It is, therefore, a hydrological norm and not a social norm. In doing so, we provide ourselves with the means to "project" a set of human actions into the physical space of the flood, making "environmentally" comparable observations made on a set of flood scenes, within the same event, or even during several events of the same nature.

The interdisciplinary analysis of extreme hydrometeorological crises through this comparison of the spatio-temporal dynamics of social actions and the phenomenon of flash floods brings about the notion of rhythmicity. The latter is based on a way of describing flood dynamics based on simple factual elements: the place and time at which the action takes place.

Starting from the idea that the origin and pace of time are not the same for everyone in the face of danger, we first describe the two simple operations necessary to set the action, on the one hand, according to the time

3 See on this point, Chapter 4 of Volume 2.

4 On the use of social networks to better understand flash flood events, see Chapter 6 of this volume.

of the flood (section 7.3), and, on the other hand, according to the pace of the flood (section 7.4). At each step, we will try to suggest, using examples, which physical cursor allows us to analyze the action during the crisis.

In the second step, we will question the physical pace that marks the occurrence of crises, their repetition in time and space. Our intention is to extend the notion of the physical cursor used for the crisis to the "pace" deemed more random between two crises. This second part is more exploratory, attempting to bring together the classical notions of risk engineering and the work presented in Chapter 5 of Volume 1 on the institutional memory of the hazard. It raises the question of the physical cursors that could be retained as predictors for returning to normals (section 7.5) and driving forces of preventive action between crises (section 7.6) before proposing a general conclusion on this notion of a cursor (section 7.7).

7.3. Flood peak as the origin of time to analyze adaptation

The question before us here is: how can we place human action in response to the flood at the time, or even the date, of the flood? To find one's bearings in relation to the flood and no longer in relation to the so-called universal time or local time implies linking the origin of time to a specific characteristic of the phenomenon that will separate the time of anticipation (before the crisis) from that of the post-crisis period.

7.3.1. *Flood peaks and peak danger*

We hypothesize that the time of the highest waters[5] is the time at which the danger is at its peak. This is objectively the moment when the flood's hold is the most significant at the scale of the watershed considered when the water levels but also the water flow velocity are the most dangerous. It is at this moment that we set the origin of the time for the event and the flood scene concerned.

It is certain that the definition of this peak danger suffers from many approximations related, on the one hand, to our ability to identify the maximum height time and, on the other hand, the hydraulic approximation

5 The peaks we are working on are related to discharges. In the case of measured quantities, they are by design unambiguously linked to water levels recorded by limnimetric stations. When it comes to modeled quantities, we maintain this hypothesis.

that is made between a high discharge and a level of dangerousness. For the first point, knowledge of discharges at all points in a river system is a hydrological problem in itself[6]. For the second point, it is the details of the local configuration that transform water discharge into danger, details that we do not take into account in our approach[7]. It is also certain that the danger may be present before and after this peak hour. We are dealing with a continuous change in terms of the situation: the rise in water, however sudden, gradually reveals the danger factors in terms of both invasion of space and force of flows. The same applies to the recession water levels that do not free up space or stop water flows instantly.

These approximations are useful because, with the help of existing measurement and modeling tools, they appear acceptable. We believe that they are negligible compared to the time adjustments required to synchronize flood scenes of the same event, which often last several hours or even more than one day (see Figure 7.1).

The lower part of Figure 7.1 shows the reaction of the main river that reached its peak during the flood on the morning of September 12 at about 7:00 a.m., 4 hours after the most intense part of the thunderstorm whose hyetogram[8] can be seen in the upper part of the figure. The reaction of the tributaries of the Gardon is represented on the intermediate graph for four size categories – two tributaries roughly dividing the basin into five (about 100 km^2) and then progressively smaller tributaries (about 50, 20, and 10 km^2).

6 Issues such as those discussed here are at the root of a hydrological problem that has recently been the subject of a scientific decade. Under the term "Predictions in Ungauged Basins" (PUB), scientists have drawn attention to the inability of traditional hydrological approaches to model the response of rivers elsewhere than where discharge measurements are made and to the need to develop methods to extend our knowledge of discharges to the entire hydrographic root [SIV 03].

7 The description of local dangerousness is the subject of work concerning, in particular, the exposure of the road network but remains difficult to address other than by a statistical approach because of the complexity of small-scale discharges that would have to be solved in a large number of scenes [NAU 13].

8 Hyetogram: in hydrology, a curve representing the intensity of rainfall according to time (www.universalis.fr).

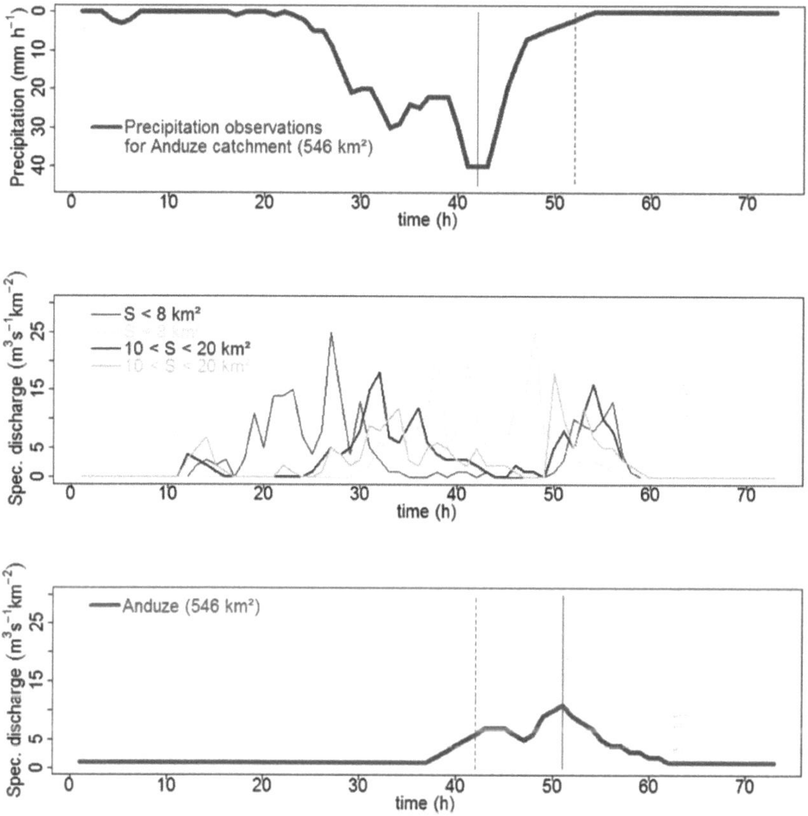

Figure 7.1. *Example of the reactions of the Gardon d'Anduze and a group of its tributaries during the catastrophic flood of September 8–9, 2002 in the Gard. The upper graph represents the average rainfall observed within the Gardon to Anduze watershed. The middle graph represents the specific discharges simulated by a hydrological model at the outlets[9] of a group of sub-basins of different areas (S). The bottom graph corresponds to the specific discharge observed at the outlet of the main basin. The red lines correspond to the peaks of rain and discharge over the same basin. Adapted from an internship report produced at LTHE. For a color version of this figure, see www.iste.co.uk/lutoff/hydrometeo2.zip*

The asynchronity of the reactions is apparent from the first level of tributaries with flood peaks at 5 and 10 a.m. (middle graph curves with the thickest lines), which frame the peak of the main river at 7 a.m. This dispersion increases with smaller tributaries (represented by increasingly

9 The outlet corresponds to the lowest point of a hydrographic network; it defines the outline of a topological watershed.

fine lines) – we can see that a basin of less than 8 km^2 reached a peak of more than 20 $m^3 \cdot s^{-1} \cdot km^{-2}$ as early as 7 p.m. the day before, even though the average rainfall intensities over the Gardon basin were still below 20 $mm \cdot h^{-1}$.

Differences in the rhythm of rising water are also apparent when considering the time required reaching the peak. This time was 2–3 hours for the main river and less than 1 hour for small tributaries.

The discharges presented were measured on the main river and modeled[10] on the tributaries. The indicated discharges are related to the surface of the watershed to indicate specific discharges (expressed in $m^3 \cdot s^{-1} \cdot km^{-2}$). The highest specific discharges were found in the smallest basins due to both the "cascade" effect that governs the spatial distribution of rainfall and the geometry of the hydrographic network [GUP 96].

7.3.2. *Multiple danger peaks in the same disaster*

The example in Figure 7.1 during the Gard disaster on September 8–9, 2002 illustrates this multiplicity of peaks. It can be observed both at a point on the river where the discharge history formed several peaks, and at different points on the hydrographic network, on the different tributaries of a river whose histories show peaks at different times.

Two hydrometeorological mechanisms are at the source of this great heterogeneity. The first concerns the propagation of water in the hydrographic network from upstream to downstream. Without being simple to represent[11], this propagation has an almost deterministic delay effect that is relatively easy to use to adapt[12]. The second mechanism is the combined

---

10 For the use of models to assess the sensitivity of hydrological responses to rainfall and soil variability in the case of the Gard in 2002, see [ANQ 10].

11 It is obvious that the propagation of water in rivers – order of magnitude of the $m \cdot s^{-1}$ – will vary considerably according to the geometry of the minor bed and, especially, of the major bed at the time of the floods of interest to us. Nevertheless, it is undoubtedly the most deterministic part of a watershed's response and underlies what we call the characteristic rapid response time of a watershed and which we relate to its size – see below.

12 For the anecdote, let us recall that there was a time when the Paillon flood warning system – a small coastal river subject to violent rapid flooding that crosses Nice – consisted in the 19th Century in spreading the warning on horseback from upstream to downstream with the

effect in space and time of the variability of rainfall inputs and soil behavior variability, which significantly complicate the estimation of river inputs[13]. It is this second mechanism that causes multiple peaks at a river point[14]. It is also this second mechanism that creates what we call flood asynchronity. This asynchronity can be defined as the deviation from "hydrological" synchronism, i.e. the phasing that flood waves would have on the hydrographic network if an effective rainfall uniformly distributed over the basin spread freely over the network [CRE 13]. As a result, asynchronicity can be summarized as what is surprising about discharges compared to what one would expect from a simple upstream–downstream logic. Figure 7.1 shows that, for several tributaries of the Gardon d'Anduze, flood peaks formed several hours after the one observed downstream in the main river.

7.3.3. *Changing the origin of time: a common adaptation issue*

Conducting an activity adapted to a fixed deadline is common. For example, it is often said that "I must do it before leaving for work" or "before moving" to indicate a time or date constraint. When the event that concerns time is shared, wanting to refer to absolute time does not bring anything, or even weakens the perception of what is constraining. It is even an opposite reasoning that leads to the definition of local times and, therefore, to the traveler's time difference. Indeed, referring to local time involves shifting the absolute time to bring it into line with events socially shared by the greatest number of people – sleep time in particular.

In the case of floods, it is, therefore, useful to situate any action that is to be analyzed in relation to the peak danger. What is confusing is to manage several floods at the same time in a small space. As we have shown above with the example of the flood in the Var department in 2011, two actors can experience a "time difference" of the peak danger when they are only a few hundred meters away.

---

cry of "Paioun ven." For the system to work, it was necessary for the "Paillon to arrive" at a speed lower than that of the horse (https://fr.wikipedia.org/wiki/Paillon).

13 Hydrologists refer to *effective* rainfall as the amount of water that will reach the river, i.e. excluding the part retained by soils and vegetation over the actual *raw* rainfall. This water contributes to the slow response of the basin which, in the case of the so-called rapid floods, represents only a minor contribution to the flood peak.

14 It should be noted that this production of multiple peaks increases toward the finest scales (see Figure 7.1).

7.3.4. *Determining peak time requires modeling*

The time origin of the flood scenes where adaptation is analyzed may sometimes have been measured when the scene is near a discharge measurement point. It is often contained in the testimonies collected from actors at the scene during the post-crisis investigation. One may wonder about the accuracy of these testimonies. It turns out that they constitute a source of information which, when brought together by hydrological modeling, has always shown that it is remarkably accurate[15]. In practice, this consistency is preferred through a model, but testimonies may be sufficient.

Strengthened by this temporal identification allowing comparisons of social responses between watersheds, how can it be used to analyze the social response process to the crisis?

7.4. The rhythm of rising water as a pace for analyzing the crisis

The definition we give of peak danger allows us to separate two periods. The forward one corresponds to the period during which we consider that the actions are anticipated – we call the anticipation time the time remaining before the peak. The post-crisis one first contains anticipatory actions that come too late in the sense of our peak danger hypothesis, but which must be analyzed as being fully part of the crisis phase (see section 7.5). It then contains, above all, the actions for a gradual return to normal (see section 7.6).

However, it is not enough to distinguish between a before and an after to analyze a sequence of actions. The urgency to act is also defined by the speed at which the danger erupts, which is related to the rhythm of the thunderstorm and the river.

7.4.1. *The pace of the thunderstorm and the river at a given point*

Figure 7.2 illustrates all the concepts mentioned above using an example of a flood observed on the Ardèche river in Sauze Saint Martin in 1993.

15 Gaume's [GAU 06] work shows how testimonies corroborate the modeling of a catastrophic flood of the Verdouble in Tautavel in 1999 – see, in particular, Figure 21 p. 37 of this report.

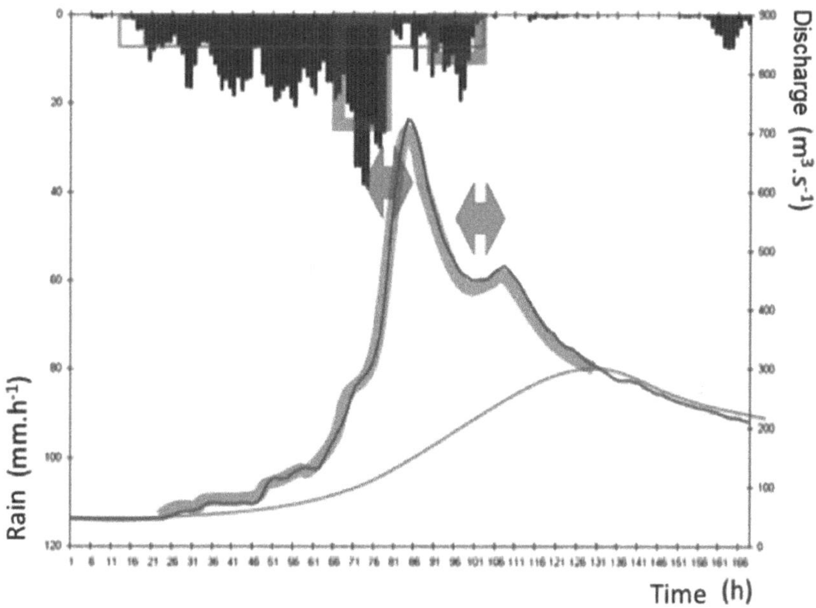

Figure 7.2. *Figure "stylizing" a flood of the Ardèche in Sauze Saint Martin (2,240 km^2) in October 1993 [ZIN 00]. The rain (upper part) and flow (lower part of the graph) chronicles are arbitrarily broken down into rapid basin response (bold lines) and slow response (fine lines)*

The first 48 hours of rainfall hardly provoked any reaction from the river, which was maintained at discharges below 100 $m^3 \cdot s^{-1}$. Then came the two series of showers that generated rapid reactions in the river – the first, the much larger one, caused the flood peak. The arrow separating the heavy rain period and the flood peak marks a basin response time of about 12 hours, close to the 14 hours indicated for this area by the formula of Marchi *et al.* [MAR 10a][16]. The response time is also characteristic of the rising water time. The duration of the showers that generated the flood peak (and the next small peak) was approximately 15 hours, illustrating quite well that the pace of weather that triggers the flood is close to that of the basin. After the flood peak, the period of water recession was limited approximately to the moment when the decrease in discharge became exponential, marking the entry into a regime dominated by groundwater recharge to the river – a gradual return to base flow via a slow response.

16 An explanation of this formula by Marchi *et al.* is given below. For this estimate of basin response times according to [MAR 10a], see also Chapter 8 of Volume 1.

This example first shows the indifference of the discharge to sustained rainfall for almost two days at the beginning of the episode. This phase can be explained by the capacity of the basin to store rain that infiltrates the soil and feeds the groundwater. These phenomena are difficult to understand due to the complexity of the environment and the physics of water transfer. They are difficult to predict and yet they determine the dangerous reaction of the basin caused by the activation of runoff phenomena that suddenly bring water to the outlet[17]. In terms of adaptation of actions, this phase fully contributes to "masking" the danger of the phenomenon – everything starts with usually heavy rain and the river is "normally" low.

This example then shows how the rise in discharge will stiffen for about 10 hours from 250 to 700 $m^3 \cdot s^{-1}$ in "response" to a series of more severe showers exceeding 20 $mm \cdot h^{-1}$ on average over the basin. The response times between the rain peak and the discharge peak are marked by horizontal arrows shown in Figure 7.2; it is approximately equal to the time of water rise.

7.4.2. *Pre-cursors (post-cursors) and risk perception*

We hypothesize that the violence of the thunderstorm and the invasion of space by rising waters are the meteorological and hydrological pre-cursors to the disaster that exposed people could use to assess the danger of their situation. The intensity of thunderstorm activity can be easily perceived by the intensity of rain and wind, electrical activity (thunder, lightning), and the decrease in daylight. Nevertheless, these indicators are difficult to quantify and are very indirectly related to each other and to flood formation. The rise in water levels is an indicator of danger that is easier to read in principle. This is obvious as the river moves to its floodplain and invades the space occupied by the person. As long as the river remains in its low-flow channel, the perception of danger is more difficult because it is finally a common situation (return time of 1–10 years[18]) and where only certain clues can signal an abnormal situation upstream to an experienced observer, such as the presence of floating objects in particular.

The cessation of rains and the recession of water are "post-cursors" of the disaster, marking the end of the dangerous situation. The cessation of rainfall is easy to observe and, in the long term, guarantees a return to normal.

17 See definition in footnote 11.

18 For more details on these periods of high water returning to the minor bed, see [WIL 78].

However, all ambiguity comes from the "term" that can often be underestimated. Downflow, or at least a decrease in the velocity of the water flow marking the end of the inflow period, is also easy to observe and is another indicator of a return to normal, although the total discharge of water from the floodplain may take longer than expected.

We propose to situate the actions carried out on each flood scene in relation to the manifestation of these pre- and post-cursors.

Figure 7.3 summarizes the time scale described above, which can be defined as the anticipation or timeliness scale[19]. The crisis time surrounding the flood peak is briefly placed with the meteorological and hydrological pre-cursor phases preceding the flood peak and the subsiding water phase. It is also possible to identify the post-crisis period, which is marked by the end of the period of rapid water level recession.

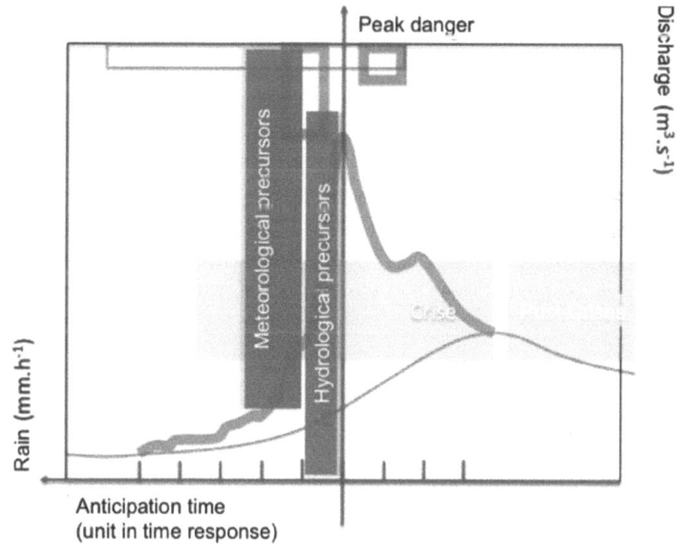

Figure 7.3. *Figure "stylizing[20]" a flood of the Ardèche in Sauze Saint Martin ($2,240\ km^2$) in October 1993: positioning on the scale of anticipation (or opportunity)*

19 This term "timeliness" is used, which contains the notion of opportunity for action in the sense that action is carried out on time [CRE 13].

20 Compared to the traditional representations used in hydrology, the detailed plots of rainfall and discharge recording have disappeared. Only their "stylized" figured figures remain.

By taking the flood peak time as the origin, and the response time of the basin as a unit, we can identify anticipatory actions carried out at the flood scene in question. It should be noted that, in this case, negative anticipation times correspond to actions taken after the danger peak (and therefore not anticipated). Hydrological pre-cursors start a unit of time before the peak, covering the period of rising water. The meteorological pre-cursors cover the preceding period, which is approximately equal in duration. It can be noted that these weather pre-cursors are less straightforward. They start relatively early, making it difficult to define the start of an event. They persist during the period of rising water. As explained in the literature (notably [CRE 13]), setting their appearance 2.6-time units before the peak, as in this graph, is an acceptable order of magnitude.

To establish the period of weather pre-cursors, we, therefore, assume that the rise in water levels is preceded by a period of heavy rain of equivalent duration. This hypothesis assumes that the times and dimensions, the characteristics of the meteorological elements (convective cell and squall line) that will trigger a sudden rise in water levels, are comparable to the response times and surfaces of the basins considered. This was empirically verified by analyzing for a set of floods the ratio between the time from the beginning of the generating rain at the peak of the flood to the time of rising water, which turned out to be 2.6 [CRE 13].

This is applicable to any basin size once the time unit is established based on the basin size itself or any other form of response time knowledge.

7.4.3. *The multiple paces of thunderstorms and floods in the same disaster*

The diversity of river responses shown in Figure 7.1 shows that the typical flood pattern described in Figure 7.2 has increasing variability from the main river to minor tributaries – i.e. smaller watersheds that, in the case of Figure 7.1, are "nestled" in the main river basin. Without quantifying things precisely, it is easy to see in this figure that the discharge increases are significantly less steep for the main basin than for its tributaries. It is also clear that the form of reactions becomes more complicated as they move to smaller scales. Figure 7.1 shows the change in rainfall only for the entire basin. Rainfall accumulations, integrated over smaller areas, will also show greater variability. This trend toward an increase in the rhythm of thunderstorm and flood growth toward small scales makes it possible to "stylize" the sequence of pre-cursors by taking, as suggested in Figure 7.3, the response time of the basin as a unit of time.

We do not describe here the hydrometeorological mechanisms that are at the source of this great diversity of responses[21]. The size of the watershed geometrically determines the time of water flow and, therefore, the time elapsed between the rain peak and the discharge peak. It explains in the first order the rate of rising water and is easy to identify for each flood scene. It can, therefore, be used as a time unit characteristic of the hydrometeorological rhythm. It is empirically shown that this unit of time is related to the size of the basin, which implicitly indicates that the propagation rate is constant at least for a range of surfaces. We propose to use the relationships established by [MAR 10a]:

$$T_L = 0.08 \cdot A^{0.55} \quad \text{for } A \leq 350 \text{ km}^2$$

$$T_L = 0.003 \cdot A^{1.10} \quad \text{for } A > 350 \text{ km}^2 \qquad [7.1]$$

where T_L[22] is the rapid response time of the basin in hours and A is its surface area in km^2, which governs the power law's change in parameters[23]. The fast response time is a good proxy for the minimum rise time and is used here to establish the pace of the river.

---

21 On this point, see Chapter 3 of Volume 1, where we have already discussed the sources of the increasing complexity of small-scale responses.

22 The response time used in [MAR 10a] is the time difference between the centroid of the rain sequence that caused the flood peak and the flood peak itself. Marchi *et al.* [MAR 10a] note that it would be more logical to consider the effective rain sequence (see definition above). We speak here of "rapid response time" insofar as the relations proposed by Marchi *et al.* [MAR 10a] are low envelopes of all the points connecting surface and response time, thus the high limit of possible propagation rates (see Figure 11 of their article).

23 Up to 350 km^2, the response time increases as the square root of the basin size – i.e. as its "equivalent length" showing that for these basins the propagation rates are the same, corresponding to flow rates on relatively steep slopes. Above 350 km^2, the hydrographic network begins to mix various propagation rates and the response time will increase in proportion to the surface, i.e. the average propagation rate will decrease with the size of the basin.

7.4.4. *Adopting the right pace: a common adaptation issue*

Conducting an activity under external constraint is common. To take an example from the field of meteorology, it is said that "I rushed home before the rain came." Studying adaptation involves bringing this modification closer to the action (accelerating walking or changing the itinerary) of the development of the external phenomenon that causes constraint (the increase of the thunderstorm). The underlying assumption is that the individual can adjust his action according to the appearance of storm pre-cursors – e.g. wind gusts and darkening of the sky. On the other hand, it is sometimes difficult to understand a change in activity if we do not know the tempo of the external constraint imposed on the person.

In the case of flash floods, the transition from a normal to a dangerous situation can take a very long time depending on where you are. It, therefore, seems necessary to take into account the pace at which this situation occurs in order to analyze how an individual or group adapts to the specific circumstances of each flood scene.

7.4.5. *Determining the characteristic pace of a place by the size of the upstream basin*

In order to analyze flood adaptation actions and make them comparable from one flood scene to another, we propose to adopt a temporal scale whose unit is the response time of the upstream hydrological basin and whose origin is the time of the flood peak. In doing so, each action can be "dated" in relation to the meteorological and hydrological precursors and characterized according to the "sequence paces" specific to each flood scene.

In the same way as the flood peak time, the hydrological rise time can sometimes be deduced from a close measurement point or from a model. In practice, we will limit ourselves to using an empirical relationship between the upstream basin surface area and the response time according to the formula of Marchi *et al.* [MAR 10a] given above. The uncertainty of this empirical approach is certainly low compared to the variability of basin sizes upstream of flood scenes, which, as we have said, cover two orders of magnitude.

Now that we have established physical benchmarks to compare the social responses and timing of actions undertaken upstream of the peak danger, depending on the watershed under consideration, let us turn to the post-crisis period.

7.5. The pace of water level recession as a post-crisis tempo

In the case of flash floods, where the rise in water levels is dominated by runoff events, the rhythm of water level recession clearly shows a dual behavior that can be used to separate the crisis from the post-crisis period.

7.5.1. *A "rapid" recession in water levels... then a more "classic" rhythm*

Flash floods occur during the rise and recession of water when rainfall supply is interrupted. Figure 7.2 shows that the hydrograph[24] is almost symmetrical with respect to the peak. The subsiding water time can thus be considered as equivalent to the rising water time. It is then logical, as shown in Figure 7.4, to keep in the crisis the brief period of rapid decline in river inflows, which remains at the same rate.

Figure 7.4 shows the main cursors of the different processes involved: meteorological processes (from the beginning to the end of the rainfall), hydrological processes (rise to flood peak level and then a subsiding in water), and social processes (crisis phase, inter-crisis, and cursors that can serve as milestones to adapt activities to the situation). The different pre- and post-cursors are to be sought in the crisis phase to determine at which stage of the hydrometeorological process we are located. These pre- and post-cursors must be differentiated from the post-event that occurs at the end of the crisis and marks the beginning of the inter-crisis period.

The cessation of rain is also often quite evident in the type of thunderstorm systems that cause flash flooding[25]. These systems, dangerous because they are stationary, often end when they return to motion and leave the affected basins. The cessation of rainfall is, therefore, a fairly safe post-cursor to the end of the crisis, provided that the time taken to transfer water into the hydrographic network is taken into account. This consideration is

24 Hydrograph: Graph showing the variation in time of some hydrological data such as stage, discharge, velocity, sediment load, and so on (hydrograph is mostly used for stage or discharge) (https://hydrologie.org/glu/EN/GF0606EN.HTM).

25 For more information on these storm systems, see Chapter 3, Volume 1.

sometimes overlooked, including by professionals, which can be the source of a dangerous announcement of an early return to normal. Let us briefly mention the example of people evacuated from a building in Remoulins during the 2002 event, due to a sudden rise in the water level of the Valliguières stream on the evening of September 8. They were allowed to return home on the morning of the 9th, the thunderstorm having moved away, before the Gard flooded them again that day[26].

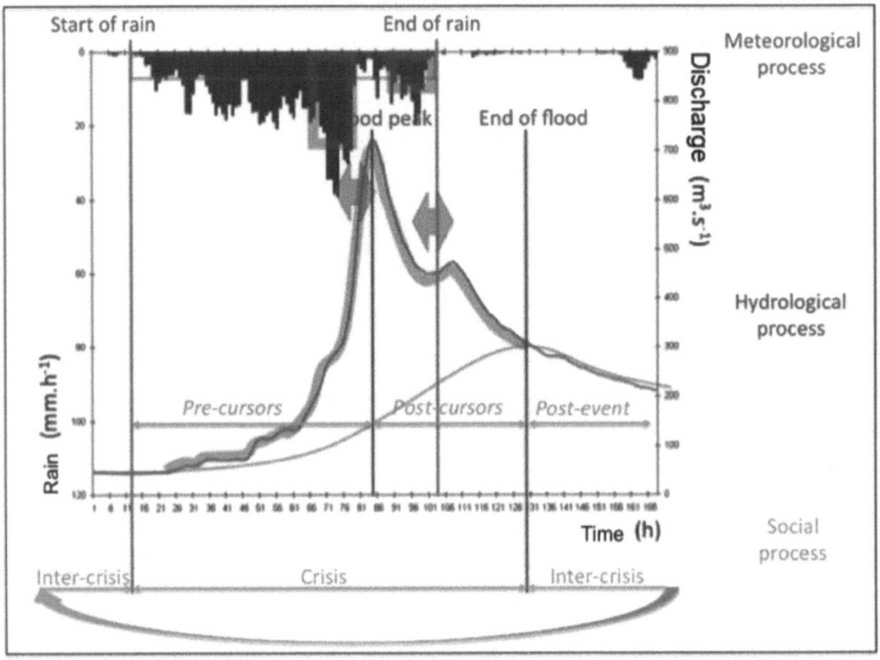

Figure 7.4. *Diagram of the socio-hydro-meteorological process phases of an isolated event*

Once the inflows by runoff are interrupted by the cessation of rain, the singularity of flash floods ends. The mechanism for controlling river flow turns into the gradual drying up of groundwater [AMB 16]. Figure 7.2 does

26 For the anecdote, let us also mention an interview carried out concerning the same disaster with a person who experienced flooding of the Alzon during the night of the 8th. He cleaned his garage on the 9th at around 5 a.m. before a new rise in water around 8 a.m. caused by the backflow of water from the nearby Gardon.

not cover a long enough period to show the slow subsiding phase of which only the beginning is visible.

The description of this subsiding water phase is only of interest in terms of a post-crisis cursor insofar as it constitutes an aggravating factor in social (prolongation of the trauma and difficulty of reorganization), material (prolonged immersion of the building in particular), and economic (e.g. paralysis of communications and transport) damage. It is particularly striking in the case of the so-called "groundwater" floods such as that of the Somme in 2001, where the duration of the receding water was measured in months.

7.5.2. *The special case of "repeated" thunderstorm events*

Observation shows that catastrophic events can occur at very short intervals of one or two days, as was the case in the Cévennes in 1958 and 2005. We can then, like earthquakes, talk about an aftershock as shown in Figure 7.5.

From a hydrological point of view, an aftershock is understood as the second thunderstorm system initiating a new overflow process by taking advantage of the state of soil saturation created by the first system[27]. The 2005 event is very representative of the particular difficulties posed by an aftershock. [MAR 09] show, for example, how the karst system characteristic of the southern Gard did not fulfill its role as a rainwater retention system. Saturated by the first event, it, on the contrary, aggravated the consequences of the second by "retro-flooding" effects. More generally, the second episode acting on saturated soils and subsoils had much more significant consequences than the first with lower cumulative rainfall. According to [MAR 10b], the weather forecast, although relevant, led to a reduction in the warning level from red to amber[28], without taking into account the state of the basins. The 2005 event highlighted the need to supplement the rainfall forecast with a hydrological forecast [VIN 07].

---

27 Figure 7.2 clearly shows how an additional, yet modest, shower causes an immediate increase in discharge within an event.

28 The French meteorological vigilance system comprises four levels, classified here in increasing order of severity: green, yellow, orange, and red.

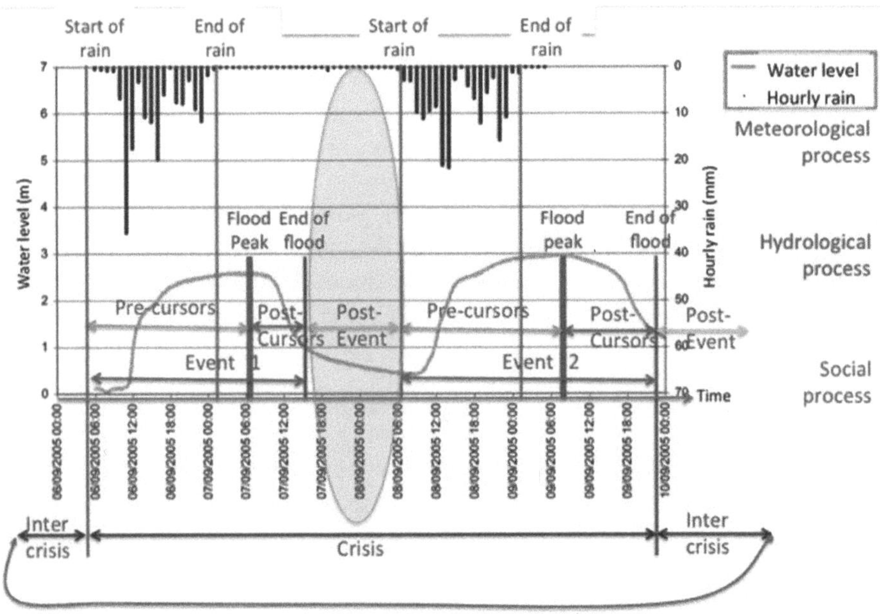

Figure 7.5. *Diagram of the phases of socio-hydro-meteorological processes: the particular case of "aftershock" events. Representation of rain intensity (meteorological process in blue), water level in the river observed between September 6 and 10, 2005 at the Rhôny outlet in Cailar [DDT 11] (hydrological process in red), and positioning of identifiable cursors for the organization of the social response (social processes in purple). The phase between the first and second event raises the problem of its definition as an indicator of a prolonged inter-crisis phase or an aftershock*

From a social point of view, the term aftershock is also appropriate in the sense that the second flood had aggravated consequences due to the absence of a restorative inter-crisis phase. For managers, the question, therefore, arose as to whether the post-event phase following the subsiding water level was the beginning of a long period of calm or an interlude before a new episode. During observation within the Gard transport management unit in 2005, Ruin [RUI 07] noted that "*this second critical phase having been less well anticipated than the previous one, we were able to observe to what extent this lack of anticipation was at the origin of disruption and stress for school transport managers*". It shows that this difficulty in qualifying the situation can have significant repercussions on transport decisions concerning the cancelation or maintenance of school transport, in particular [RUI 07]. The exploration of these short inter-crises is, therefore, a major challenge in the management of these phenomena, an issue all the more significant since fall 2014 showed that aftershocks over short periods of violent events in the same region was no exception. Indeed, such sequences were observed on a recurrent basis between 15 September and 15 October throughout the south-eastern region of France [PAY 15]. A better understanding of the tempo and characteristics of these phenomena from both meteorological and hydrological points of view is, therefore, necessary to better qualify this critical moment, the post-event and inter-crisis phase.

7.6. The rate of occurrence of floods as a tempo for analyzing the inter-crisis period

The pace of the inter-crisis period is marked by that of the occurrence of crises, i.e. by the occurrence of “high-impact” events (see Chapter 4 of Volume 1). These events stimulate institutional action at both local and national levels (see Chapter 5 of Volume 1). The question is whether the pace of these “high-impact” events can be characterized by a statistical or physical approach. It would then be possible, as it was the case during the crisis, to compare these characteristic paces with that of institutional action analyzed at a given territorial scale. In the same way that we have seen the link between the rate of rising water and the size of watersheds, we are invited here to consider a possible link between the size of an institutional territory and the rhythm of the “high-impact events” that have marked it. This link could be used to explain, for example, a difference in institutional tempo between two regions or two countries. We are thinking, for example, of Great Britain and France. In their article, [PEN 06] highlights a characteristic 30-year time frame for scoring the evolutionary stages of flood legislation from the influence of the land reclamation lobby to the influence of dikes and then zoning. In Chapter 5 of Volume 1, we showed that this characteristic time in France is rather 70 years. There are, of course, many political and social reasons for this difference. However, it remains interesting to question the nature and frequency of catastrophic events that could “accelerate” legislative change (to use the concept proposed in [PEN 06]).

7.6.1. *The probability of complex events within a territory*

Human exposure, and therefore risk management, is locally tested with return periods ranging from 10 to 100 years, depending on the vulnerability level of the system. In France, urban rainwater sanitation facilities are designed for a 10-year thunderstorm, dikes are designed for 100 years, and dams for an exceptional flood with a return period of 10,000 years. The probability of occurrence of rainfall levels or discharges is calculated in a traditional way using data relating to a point – rain at a point in the territory, discharge at the outlet of a watershed[29], on which observations are available.

However, risk exposure cannot be reduced to a one-time vision. The institutional response concerns interlocking territories: a meteorological or hydrological warning given at the departmental level applies uniformly to all the municipalities in the department. To be more precise, this institutional response must, therefore, be informed by probability calculations giving the frequency of events affecting this nesting of territories.

As shown in the previous sections, flash floods are complex events in the sense that they can combine significantly different elementary events. Many French conurbations along rivers such as Paris, Bordeaux, or Grenoble are exposed to the co-occurrence of very localized thunderstorms over the city while the main river is flooded or subjected to the tide. They are also complex in the sense that they can concern a large territory. We showed in Chapter 5 of Volume 1 that French legislation has often reacted to flood events that are widespread in a large part of the country. This was the case, for example, of the 1856 floods that affected both the Loire and the Rhône or those of 1930 that affected both the Garonne and the Languedoc. In some cases, events that trigger a legislative response associate two very different forms of flooding but on the same scale, for example, in 2010, with the marine submersion in Charente and the Draguignan flash flood.

Probabilizing such events are, therefore, necessary to help analyze territorial and social sensitivity. Defining the probability of high-impact events would allow for a better analysis of the rates of institutional responses at different scales. What are the characteristics of events or sequence of events that trigger a local, regional, national response? Answering this question and giving oneself the physical reference cursors would then make it possible, as for the crisis, to give a more precise pace to the between crises. But how is this probabilization done? What is the link with large atmospheric forcings and seasonal forecasting?

29 The idea of linking rain and discharge in a probabilistic approach comes from EDF engineers Guillot and Duband, who proposed, in the 1960s, a radically new approach to estimating long return period discharges – the Gradex method [GUI 67].

7.6.2. *The pace of major atmospheric forcings and seasonal forecasting*

The occurrence of intense and quasi-stationary rainfall events is often linked to the exceptional coincidence of several meteorological factors acting at different scales [DUC 08], [NUI 08], [RIC 12].

On a continental or even global scale, the positioning of low- and high-pressure zones determines the direction and amplitude of atmospheric flows. These positions generally evolve quite slowly and are characterized by "oscillation indices," such as ENSO (El Nino Southern Oscillation) for the Southern Hemisphere and NAO (North Atlantic Oscillation) for Western Europe for example. They lead to different weather patterns (or "weather types") depending on the values of these indices. In the case of the north-western Mediterranean, quasi-stationary intense rainfall events are very often linked to the presence of a low-pressure area over the nearby Atlantic (or Iberian Peninsula) that generates warm and most air flows from the Mediterranean Sea on one side and cold and drier air flows on the other. However, this configuration is not sufficient to explain the onset of extreme events. Other factors with different time frames are required. The temperature of the surface of the Mediterranean Sea is one of them; it has a slower (annual) variability and a more or less marked amplitude depending on past meteorological situations (the sea "stores" energy that it can release several months later). The convergence of air flows in the lower layers of the atmosphere – often explained by the circumvention of landforms – is another; it acts on a more local scale and its occurrence in the same place is much rarer.

The concomitance of different factors such as those mentioned above, each of which has its own pace, can lead to violent phenomena because they trigger the formation of heavy precipitation systems with relatively little mobility. This near-stationarity sometimes gives the ability for already formed systems to maintain themselves. Quasi-stationarity is also a matter of pace: systems move slowly, or even stagnate, when the weather pattern is frozen or evolves very slowly. It should also be noted that other "blocking" situations, such as the persistence of a high-pressure zone, are at the origin of other high-impact events such as cold waves, heat waves, and/or periods of high air pollution.

Weather forecasting models now allow us to estimate trends for the coming weeks and months, taking into account, for example, feedback between the atmosphere and the ocean. It is hoped that this will predict future weather patterns and their consequences on the local, regional, or national occurrence of an "extreme rain event," which, if it resonates with the hydrological system, will turn (or not) into a catastrophic flood. Thus, if the emergence of these highly unfavorable patterns was expected and the links with the flood were established, then we could talk about seasonal pre-cursors whose occurrence could be used to characterize the periods between crises. However, it would also be necessary to be able to predict the probability of concomitant large-scale configurations with other more local factors whose predictability is still very low today [KUN 05], [MUL 03].

7.7. To conclude: interpreting the cursors

While articulation in physical and social processes is necessary to understand how to respond to and anticipate extreme hydrometeorological events, its implementation requires specific interdisciplinary knowledge developments. By interdisciplinary knowledge, we aim for knowledge that may be more or less basic for a discipline (such as flood peak and watershed response time), but which, when combined with another discipline, changes the way we look at the processes studied (analysis of the timing of social responses according to the tempo imposed by the flood).

This chapter proposes an interdisciplinary interpretation of the physical cursors taken as a reduction of hydrometeorological reality that is sufficiently simplistic enough to be introduced into the analysis of the action and is sufficiently informative to accurately reflect the environmental conditions of this action. This Physical-social study thus proposes a temporal scale centered (on the flood peak) and reduced (by response time or by recurrence) which makes it possible to jointly analyze physical and social cursors during the crisis and between crises. These cursors express the climatic occurrence of extreme hydrometeorological events over a set of watersheds that react asynchronously to a common meteorological event. This standardization process, after having served as a material for interdisciplinary collaboration, in turn raises new questions specific to each discipline.

From the point of view of the social sciences, this standardization of observations by physics allows a shift in reference factors that renews the analysis of social responses to flooding. While the more traditional differentiation criteria, such as age, gender, etc., remain relevant in a second stage of the analysis, physical cursors make it possible first to focus on the rates of individual or collective reaction and adaptation to an environmental constraint that is imposed in a differentiated way according to the spatial and temporal scales considered.

From the point of view of climate science, one might have thought that meteorological forecasting, a skillful interpretation of physical precursors, would be sufficient to provide the necessary references to normalize social observations. However, attempts in this direction (e.g. [RUI 04]) quickly indicate limits. The question of the scale at which this forecast is provided (so far the departmental level) is not sufficient to understand the actions undertaken more locally by individuals. This is the view of many local decision makers for whom the levels of vigilance proposed by Météo France, which should normally be "official" precursors to an event, only serve in part to mobilize the social response, both at the individual and collective levels (on this point, see [DEM 19]). Nevertheless, work in progress should allow a more accurate spatial forecast that could then eventually serve as a real precursor to organize social response at the local level.

In the inter-crisis approach, we note the current use of flood markers as a hydrological cursor, associating a water level on a specific point of the territory with a past event. If they are mainly used as communication and awareness-raising tools today, we could consider their possible mobilization to better understand the prevention processes in the territories under consideration. Have the events identified locally by these cursors led to changes in the way risk is managed or prevented? But also, by moving to a broader scale, what are the events noticed via these markers, what can we say about their pace, and how did they stimulate or not stimulate preventive responses through the different interlocking scales (locally, regionally, and nationally)? Answering these questions implies pursuing this socio-hydro-meteorological approach and enriching the interdisciplinary toolbox by crossing approaches.

7.8. References

[AMB 16] AMBROISE B., "Variable water-saturated areas and streamflow generation in the small Ringelbach catchment (Vosges Mountains, France): the master recession curve as an equilibrium curve for interactions between atmosphere, surface and ground waters", *Hydrological Processes*, vol. 30, pp. 3560–3577, 2016.

[ANQ 10] ANQUETIN S., BRAUD I., VANNIER O. *et al.*, "Sensitivity of the hydrological response to the variability of rainfall fields and soils for the Gard 2002 flash-flood event", *Journal of Hydrology*, vol. 394, no. 1, pp. 134–147, 2010.

[CRE 13] CREUTIN J.D., BORGA M., GRUNTFEST E. *et al.*, "A space and time framework for analyzing human anticipation of flash floods", *Journal of Hydrology*, vol. 482, pp. 14–24, 2013.

[DDT 11] Direction départementale des territoires et de la mer du Gard, BRL ingénierie, Plan de prévention des risques inondation haut Vistre, Buffalon et moyen Vistre. Phase 1: Analyse du fonctionnement du bassin versant et de ses principaux enjeux, 162 p., 2011.

[DED 07] DEDIEU F., La sécurité civile dans la tempête: autopsie organisationnelle de la catastrophe du 27 décembre 1999, Thesis, IEP Paris, Paris, 2007.

[DED 15] DEDIEU F., *Une catastrophe ordinaire: la tempête du 27 décembre 1999*, Éditions de l'École des hautes études en sciences sociales, Paris, 2015.

[DEM 19] DEMERITT D., STEPHENS E.M., CRÉTON-CAZANAVE L. *et al.*, "Communicating and using ensemble flood forecasts in flood incident management: lessons from social science", in DUAN Q., PAPPENBERGER F., WOOD A. *et al.* (eds), *Handbook of Hydrometeorological Ensemble Forecasting*, Springer, Berlin, Heidelberg, 2019.

[DOU 17] DOUVINET J., GISCLARD B., KOUADIO J.S. *et al.*, "Une place pour les technologies smartphones et les Réseaux Sociaux Numériques (RSN) dans les dispositifs institutionnels de l'alerte aux inondations en France?", *Cybergeo: European Journal of Geography*, vol. 801, 2017, available at: https://journals.openedition.org/cybergeo/27875.

[DUC 08] DUCROCQ V., NUISSIER O., RICARD D. *et al.*, "A numerical study of three catastrophic precipitating events over southern France. II: mesoscale triggering and stationarity factors", *Quarterly Journal of the Royal Meteorological Society*, vol. 134, pp. 131–145, 2008.

[GAU 06] GAUME E., Post Flash-flood investigation – methodological note, Report D23.2 from the Floodsite project, available: http://www.floodsite.net/, 62 p., 2006.

[GUI 67] GUILLOT P., DUBAND D., "La méthode du GRADEX pour le calcul de la probabilité des crues à partir des pluies", in *Colloque International sur les crues et leur évaluation, Leningrad août 1967*, pp. 15–22, 1967.

[GUP 96] GUPTA V.K., CASTRO S.L., OVER T.M., "On scaling exponents of spatial peak flows from rainfall and river network geometry", *Journal of Hydrology*, vol. 187, pp. 81–104, 1996.

[KUN 05] KUNDZEWICZ Z.W., ULBRICH U., GRACZYK D. *et al.*, "Summer floods in Central Europe–climate change track?", *Natural Hazards*, vol. 36, nos 1–2, pp. 165–189, 2005.

[LIN 11] LINDSAY B.R., Social media and disasters: current uses, future options, and policy considerations, Congressional Research Service, 7-5700, R41987, 2011.

[MAR 97] MARTI R., LEPELLETIER T., "L'hydrologie de la crue de 1910 et autres grandes crues du bassin de la Seine", *La Houille Blanche*, vol. 8, pp. 33–39, 1997.

[MAR 09] MARECHAL J.C., LADOUCHE B., DÖRFLIGER N., "Analyse hydrogéologique de la contribution de l'eau souterraine à la crue éclair des 6 et 8 septembre 2005 à Nîmes", *La Houille Blanche*, vol. 2, pp. 88–93, 2009.

[MAR 10a] MARCHI L., BORGA M., PRECISO E. *et al.*, "Characterisation of selected extreme flash floods in Europe and implications for flood risk management", *Journal of Hydrology*, vol. 394, pp. 118–133, 2010.

[MAR 10b] MARTY R., Hydrological ensemble forecasts for flash flood basins, PhD thesis, Université de Grenoble, Grenoble, France, 2010.

[MEL 97] MELUCCI A., "Rythmes internes et rythmes sociaux dans un monde planétaire", *Nouvelles pratiques sociales*, vol. 10, pp. 195–202, 1997.

[MUL 03] MUDELSEE M., BÖRNGEN M., TETZLAFF G. *et al.*, "No upward trends in the occurrence of extreme floods in central Europe", *Nature*, vol. 425, no. 6954, pp. 166–169, 2003.

[NAU 13] NAULIN J.P., PAYRASTRE O., GAUME E., "Spatially distributed flood forecasting in flash flood prone areas: application to road network supervision in Southern France", *Journal of Hydrology*, vol. 486, pp. 88–99, 2013.

[NUI 08] NUISSIER O., DUCROCQ V., RICARD R. *et al.*, "A numerical study of three catastrophic precipitating events over southern France. I: numerical framework and synoptic ingredients", *Quarterly Journal of the Royal Meteorological Society*, vol. 134, pp. 111–130, 2008.

[PAY 15] PAYRASTRE, O., Estimation des débits de pointe des crues de l'automne 2014 dans le Gard et l'Hérault, Convention DGPR-Ifsttar 2014 no. 2101388232, September 10, 2014, Action 4: Appui au SCHAPI, 2015.

[PEN 06] PENNING-ROWSELL E., JOHNSON C., TUNSTALL S., "'Signals' from pre-crisis discourse: lessons from UK flooding for global environmental policy change?", *Global Environmental Change*, vol. 16, pp. 323–339, 2006.

[RIC 12] RICARD D., DUCROCQ V., AUGER L., "A climatology of the mesoscale environment associated with heavily precipitating events over a Northwestern Mediterranean area", *Journal of Applied Meteorology and Climatology*, vol. 51, pp. 468–488, 2012.

[RUI 04] RUIN I., LUTOFF C., "Vulnérabilité face aux crues rapides et mobilité des populations en temps de crise", *La houille blanche*, vol. 6, no. 6, pp. 114–119, 2004.

[RUI 07] RUIN I., Conduite à contre-courant. Les pratiques de mobilité dans le Gard: facteur de vulnérabilité aux crues rapides, PhD thesis, Université Joseph-Fourier-Grenoble I, Grenoble, France, 2007.

[RUI 14] RUIN I., LUTOFF C., BOUDEVILLAIN B. *et al.*, "Social and hydrological responses to extreme precipitations: an interdisciplinary strategy for post-flood investigation", *Weather, Climate and Society*, vol. 6, no. 1, pp. 135–153, 2014.

[SIV 03] SIVAPALAN M., TAKEUCHI K., FRANKS S.W. *et al.*, "IAHS decade on predictions in ungauged basins (PUB), 2003–2012: shaping an exciting future for the hydrological sciences", *Hydrological Sciences Journal*, vol. 48, no. 6, pp. 857–880, 2003.

[VIN 07] VINET F. "Approches nationales de la prévention des risques et besoins locaux: le cas de la prévision et de l'alerte aux crues dans le midi méditerranéen", *Géocarrefour*, vol. 82, nos 1–2, pp. 35–42, 2007.

[WIL 78] WILLIAMS G.P., "Bankfull discharge of rivers", *Water Resources Research*, vol. 14, no. 6, pp. 1141–1154, 1978.

[ZIN 00] ZIN I., SAULNIER G.-M., "Assessing ambiguity in physically-based hydrological models: stochastic approach with application to the Ardeche catchment (Southern France)", *Accuracy 2000*, Amsterdam, The Netherlands, pp. 759–766, 2000.

8

Method for Observing the Rates of Exposure to Flash Floods: Physical and Social Processes

8.1. Introduction

Observing exposure to flash floods requires a joint description of the mobility of water and people during crises and also between crises (see Chapter 8 of Volume 1). The observation concerns a sufficiently large group of rivers and a sufficiently long period of time to be able to have a sufficient number of events, which by definition are rare and localized. It is conducted by a balanced combination of skills, scientific or operational objectives, and observation resources from both geosciences and the human and social sciences.

The observation resonates with questions such as:

– What are the phenomena that contribute to exposure to flash floods?

– What are the scales from the short term of the crisis to the long-term of the inter-crisis that need to be investigated?

– What quantitative and qualitative, elementary or integrated, and direct or indirect data should be used to describe these phenomena?

Chapter written by Céline LUTOFF, Séverine DURAND and Jean-Dominique CREUTIN, as well as all other contributors to the ANR Mobiclimex Project: Isabelle ANDRÉ-POYAUD, Sandrine ANQUETIN, Joël BOULIER, Isabelle BRAUD, Edwige DUBOS-PAILLARD, Véronique DUCROCQ, Sylvie DUVILLARD, Antonin PAVARD, Isabelle RUIN, Valérie SANSEVERINO-GODFRIN and Béatrice VINCENDON.

The variety of phenomena and scales immediately places them in a multidimensional problem – time and space, and also those that describe more specific quantities such as levels of social connections and institutional or management functions (see Chapter 1 of Volume 1).

The exercise carried out in this chapter puts a set of physical and human phenomena contributing to flood exposure against data available to analyze these phenomena individually, but especially as they cross and interact.

The common theme proposed for this comparison is that of scales: what scales characterize the analysis of exposure rates during the crisis and in the inter-crisis? The idea is to characterize all the data used by temporal and spatial resolutions and to show their adequacy for the analysis of the phenomena governing exposure. This logic is unequally shared or even discussed by the disciplinary panel necessary for the study of exposure. Time is used in various forms; unlike time, space is not a dimension that can be easily used by everyone.

This chapter begins with a brief description of a data set, covering several aspects of exposure to flash floods in France[1] (section 8.2). Then, we show the nature, strengths, and weaknesses of the observation methods mentioned (section 8.3). Finally, we return to the description shared between physical and human data and show how it leads to a hierarchy of the phenomena contributing to exposure and what are the scales useful for this description (section 8.4).

8.2. Comparison of phenomena and data

The data description exercise is taken as a means of breaking down the interaction process between water mobility and human mobility into elementary physical and human phenomena[2], in the sense that they are understood by describing the means available to document them.

1 These are, in fact, the aspects addressed by the MobiClimex Project (see Foreword) whose work feeds into this book.

2 The word phenomenon is familiar in physics (we will talk here about the meteorological and hydrological phenomena generating the flood). It is also understandable in the human sciences if we admit that "moral facts are phenomena like any other; they consist of rules of action that can be recognized by certain distinctive characteristics; it must, therefore, be possible to observe, describe, classify, and search for the laws that explain them" (Durkheim,

All the phenomena and data discussed in this chapter are summarized in Table 8.1 (in Appendix 2). We describe flood exposure from the time of the crisis (short term) to the time between crises (long term)[3]. To complete Table 8.1 (and to avoid the reader having to refer to it too often during reading), we indicate, in footnotes, for each documented phenomenon, the temporal and spatial resolutions of the data used. The aim is to yield orders of magnitude that lead to reflection on the approach of each phenomenon and ultimately make it possible to assess the homogeneity of the overall approach.

8.2.1. *Reconstructing daily traffic on the Gard's roads*

Over the short term, people's mobility is described as the articulation of a daily program of activity (temporal course of action) and a movement (spatial progression of individuals as they carry out their activity). The study of daily activity programs is a very active research field in relation to the organization of the city, transport (since the work of the sociologist Chapin [CHA 74] and that of the geographer Hägerstrand [HAG 70] who initiated research in "Time Geography"), and, more recently, in relation to climate events and natural risks [BOC 13], [COO 13], [KIL 07]. In Chapter 5, Shabou *et al.* rely on surveys of sufficiently large[4] samples of people to then statistically construct standard time use profiles by socio-economic categories (see section 8.5). These profiles contain information on people's mobility – lengths, durations[5], schedules, and modes of travel – and, in the background, on their places of immobility. Giving a spatial dimension to this temporal information requires a geographical knowledge of places of mobility and immobility – mainly roads and buildings – and of the distribution of people in these places. Census data provide statistical

Work. Div., 1893, according to the lexicography of the Centre National de Ressources Textuelles et Lexicales – http://www.cnrtl.fr). We maintain the term process to refer to a set of phenomena that contribute to an outcome such as human mobility or, more generally, exposure to floods. We simply find it interesting to note here the possibility of breaking down the complex processes that we study with a more or less significant level of detail.

3 It is interesting to note that this division between the short and long term is thematically as strong as the one between water and people's mobility. Whether in the physical or human domain, thematic skills, data, and methods differ significantly depending on which one is interested in the short or long term – dealing with the entire exposure process, thus requires the involvement of at least four scientific communities.

4 Spatial resolution taken at the size of the department (77 km for the Gard whose area is 5853 km^2) in which the sample of people surveyed is dispersed.

5 Temporal resolution typically of 1 minute, depending on the survey mode.

information on the municipalities[6] of residence and work or education of the entire population[7]. The location of tax revenues by parcel of land indicates the dispersion of places of residence and commercial and industrial activity within municipalities[8]. Road maps[9] give an idea of the possible routes within and between municipalities to connect these different places. The "reconstruction" of these paths can only be statistical and made under several assumptions. It is statistical insofar as it is the socio-demographic profiles of the municipalities' inhabitants that make it possible to distribute geographically the activity of the entire population. It is based on assumptions, in particular, about the spatial distribution of people's activities in the construction of municipalities and their choice of route (see Chapter 4). Traffic counts make it possible to validate at a few points all the daily flows thus reconstructed[10].

8.2.2. *Describing the floodability of the road*

In a very general way, over a short period of time, the flooding is seen as the submergence of the territory by surface runoff resulting from excessive precipitation, either cumulatively or intensively (see Chapter 3 of Volume 1). At the hydrometeorological level, observations of rainfall and river flows are available in an operational way[11]. The modeling of the transformation of rain into river flow makes it possible to reconstruct, particularly at ungauged points, flow chronicles at all points in the hydrographic network[12] [BRA 18]. One difficulty is then to transform these

6 Spatial resolution taken at the size of the municipalities (in the Gard the average surface area of the municipalities is 16 km^2).

7 The origin-destination matrices derived from the census indicate the flows linked to home-work journeys within (diagonal of the matrix) and between the municipalities of a department and neighboring departments.

8 Spatial resolution of 100 m in relation to the average plot size.

9 Spatial resolution of the road size, e.g. 10 m.

10 The proposed traffic reconstitution does not allow at least two classes of problems to be taken into account: the variety of modal choices (in particular, the number of people per private or public transport vehicle) and the traffic of road professionals (in particular, transport operators) [FOU 12].

11 The temporal resolution of these data is in fractions of an hour (from 15 minutes to an hour). Their spatial resolution is now better for rain (1 km^2 is typical of radar data) than for flow (at best 10 km^2 for the surface of the smallest instrumented watersheds – the majority of them being one or two orders of magnitude above).

12 The model results have a typical resolution similar to that of the data.

flows into a flood field to highlight damaging overflows. We focus here on the submergence of roads by water at the intersection of the hydrographic network and the road network[13]. The Gard department has a database, describing the road points regularly cut off by floods. This knowledge has enabled learning about automatic recognition of such points and their sensitivity to the flooding by crossing general topography, river geometry and road geometry[14] [NAU 12], [VER 07]. This knowledge of the sensitivity of the cross-road network to the modeling of a particular storm event provides a probabilistic knowledge of the road flooding. It is probabilistic insofar as the sensitivity of the cut-off points is defined in a probabilistic and uncertain way – it is difficult to give a deterministic relationship between a water level in the river and the flooding of a road due to the local complexity of the flows[15].

8.2.3. *Explore the effect of individual decisions on population exposure*

The knowledge of daily movement crossed with the knowledge of cut-off points provides a potential exposure that individual decisions (to adapt to flooding or not) will change. Behavioral patterns appeared in the interviews, always in limited numbers, based on feedback[16]. It is possible to test the effect of behavioral scenarios on overall exposure by calculation. A decision model links the perception of a deterioration in environmental circumstances (rain and light), traffic difficulties (submerged section of road), and information received (alert level) to a decision: continuing without adaptation, stopping at a cut-off point, delaying the next activity, and

---

13 This vision is reductive in the sense that it favors mobile exposure. It must be completed by an analysis of immobile exposure (in buildings, residential, or various activities). However, reaching these buildings requires roads and the two categories of exposure are, therefore, closely linked.

14 The spatial resolutions of these data are decametric and their temporal resolution is that of a decade or even a century.

15 This includes the complex geometry of the areas concerned (bridges, dikes, etc.), their possible modification by the force of the flows, as well as the presence of objects in the flow (ice jams). This limit is true for roads but for any other part of a territory when the flows are violent and loaded.

16 Chapter 4 of Volume 2 discusses the verified accuracy of the physical information on the flood given by witnesses, even several months after the event. We can talk about the resolution at 5 minutes/m. It is clear that in terms of behavior, the question of "precision" depends on multiple psychological mechanisms.

returning home. This model balances these external circumstances against the personal constraints of the planned travel program[17] [SHA 16].

8.2.4. Extreme events, impact and regulatory developments

Over time, the flooding is first seen as a phenomenon statistically summarized by the occurrence of extreme daily rainfall accumulations. This knowledge, insufficient to identify high-impact events, is complemented by the historical compilation of dates when such rainy events have been recognized by their human and material impacts[18].

The impact of extreme events can also be investigated by the evolution of institutional tools for space and risk management. This evolution concerns different institutional levels: the individual seeking a building permit, the municipality and its urban planning and risk plans and, finally, the State and its legislative and regulatory role. The data used concern these three levels[19], and the exercise consists in examining whether the rate of occurrence of crises changes the institutional dimension and how different levels interact. At the individual level, the impact can be analyzed according to the rate of issuance, and the characteristics of building permit with regard to floodability (see Chapter 2 of this volume). At the communal and national levels, the legislation is examined with regard to risk and its local application through the implementation of urban planning and risk management plans (see Chapters 5 and 6 of Volume 1).

17 The resolutions of the meteorological and road data used are the same as those seen above. Météo France's alert levels have a departmental granularity and a time display. The resolution of the travel programs is seen above.

18 The impacts are marked in the recent period in France by the natural disaster decrees that are issued at municipal level; hence, the spatial resolution of a few kilometers is at the average size of the municipalities in the Gard.

19 The spatial resolution of these tools can be located either between the meter and the parcel, based on the resolution of cadastral maps, or at the size of the municipality, because the evolution of the tools is partly the responsibility of the municipalities, and, finally, at the size of the State. The rate of this evolution is from "annually" to "every 10 years".

8.2.5. *Regional residential mobility and its link to flooding*

Residential mobility is documented at two levels (that of the individual and buildings and that of the communes) and on the basis of two ways of observing – the monitoring of trajectories or of states, i.e. stocks.

Individual trajectories of residential migration can be collected directly through narratives – from the individual on his or her own trajectory or that of residential mobility professionals such as real estate agents or lawyers[20]. The collection of these accounts also makes it possible to qualify the relative influence of the floodability of sites on residential choices.

Residential trajectories can also be assessed statistically using specific surveys on the motivations for moving and census data related to the previous residence. The granularity and nature of the information are no longer the same – apart from not being able to qualify the reasons for these changes, the intra-communal migration is not seen and the pace is that of censuses, the profiles are types and not real examples – but the result reflects global trends at population level.

Regular mapping of the territory is a way of monitoring the decision to build in a particular space and assessing its behavior with regard to flooding. The comparison of successive maps[21] allows us to perceive movement; the knowledge of floodplains places it in terms of exposure. The temporal resolution of this approach can be refined by taking into consideration the tax data attached to each parcel. In particular, these data give an indication of the construction date (building permit application or simple declaration), which improves the resolution by an order of magnitude and, above all, considerably extends the period covered[22]. It should be noted that the decision to build reflects that of the people if we accept an effective or, more generally, homogeneous occupation in the buildings' time and space.

20 Whether for the individuals concerned or professionals, the spatial resolution is fine, at least that of the neighborhood, and the temporal resolution is typically annual.

21 The accuracy of the maps is metric; their publication rate is 10 years. Advances in aerial and spatial imagery have improved these characteristics in recent times.

22 The declaration of a plot of land degrades the resolution of the positioning of the buildings, particularly with regard to the floodplain. The uncertainty over the year is linked to the delay between the building permit application and construction, or even to errors in the declaration of the age of the buildings.

At the departmental level, previous observations make it possible to describe the exposure of all municipalities by counting their buildings in flood risk areas. We can also obtain a set of intermunicipal flows[23] based on the apparent migratory balances of the communes and the knowledge of standard residential mobility profiles.

8.3. The main classes of data: their contributions and limitations

Among the set of observations described in the previous section, we focus here on their sampling characteristics, their direct or indirect nature, and their level of integration. This analysis aims to explore the methodological homogeneity of observing a problem that combines physical and human sciences.

8.3.1. *Sampling in time and space – resolution and field of observation*

The concepts of temporal and spatial resolutions are used in Table 8.1 to characterize the data sampling. It is relevant to define the elementary steps of time and space of observation, on the one hand, to better understand what the data are and, on the other hand, to judge their adequacy to what is known about the dynamics of the phenomenon. The sampling is also characterized by the notion of the domain of study which indicates the spatial and temporal extents covered by the data. Reading Table 8.1 with regard to these concepts calls for several remarks that show how these concepts usefully cross physical and human phenomena and their intersection.

The concepts of resolution and time domain are probably the easiest to share and cover more or less regular sampling modes. This can be referred to as sampling frequency, i.e. continuous sampling at a regular time step over long periods of time for the meteorological and hydrological data set on water mobility[24]. Perhaps more unexpectedly, we can also speak about the

23 These movements concern decennial resolution. They reflect the major changes in the territory.

24 Flash floods have prompted a specific effort to observe rain and flows with a resolution of one or two orders of magnitude finer than that of conventional hydrometeorological observation (such as that of weather radar data, in particular) – see Chapter 1 of Volume 1 and Chapter 7 of this volume, as well as Borga *et al.* [BOR 08]. As this effort is recent, the

sampling frequency with the regular publication of the *Journal Officiel de la République*, which makes it possible to observe the State's institutional dynamics. Over a very long period of time, this publication gives access to the construction of the law with a daily resolution that seems redundant in view of the pace of its evolution when we look at a given field[25]. Similarly, the regularity of administrative mechanisms – annual for taxation (income, assets, including construction) or every 10 year for the census and land register – making it possible to monitor the construction or intermunicipal migration on a regular basis but, in many respects, too loosely (see Chapter 2 of this volume).

With regard, more specifically, to flooding, submersion, and crisis, only occasional observations are available, such as higher water maps used to identify floodplains or the PITCH database, created by the Gard Roads Department, which describes the floodability of the most exposed road sectors. Here, we find the great difficulty in characterizing the observation of extremes and the possible confusion between the almost instantaneous time of the measurement, its consolidation over the duration of an event (post-flood survey on the highest observed waters) or over the duration of an observation period (several years for PITCH), and its attribution to a return period.

The notion of the spatial resolution is, therefore, more difficult to use because it must cover very diverse modes of the sampling space.

The only regular grids in space are probably those concerning imaging – weather radar for rain, aerial photography, or satellite for maps. Their resolution is given by the physical characteristics of the sensor, or even by its data acquisition system (per kilometer for radar). Several phenomena are due to an irregular segmentation of space (buildings, parcels, municipalities, and watersheds) or a branch structure (hydrographic and road networks). Their spatial resolution can be taken as that of the method used to establish the map that represents them (ground survey, aerial photography – 1 m or less) or as that of the object that we want to "solve" (decametric for the road, river, or

temporal domain of the data lacks depth and forces us, in order to have long series, to return to the classical data, therefore less well resolved in space and time (such as rain gauge data).

25 Chapter 5 of Volume 1 shows a characteristic time of 70 years for the evolution of the law with regard to flooding in France.

building, hectometric for the plot, municipality[26], etc.). Buildings, roads, and all cadastral information on property and even floodability have benefited, with digitization, from enormous progress in terms of spatial resolution and homogeneity [CRO 16], [KAU 98]. Regulatory phenomena can be considered resolved by definition by the size of the entity to which they apply.

Finally, it is interesting to see whether the notion of resolution, temporal or spatial, applies to a survey such as national transport and travel surveys or to an interview as seen with real estate professionals or after crises with individuals. The daily activity programs and their possible disruption by flooding, on the one hand, and the residential trajectories, on the other hand, are an initial division of individuals' activity between high and low frequency. If this breakdown makes sense[27], the temporal resolutions attached to these two phenomena are defined by the interviews or surveys used. The form of the questions and the nature of the activities they cover – safety during the crisis or moving during the inter-crisis period for example – lead respondents to spontaneously choose fairly homogeneous temporal resolutions – minute by hour for the daily and monthly to annual for life trajectories[28].

8.3.2. *Nature and level of integration of observations*

We will examine whether the data come from a direct observation of the phenomenon or whether they are derived from observations of-related phenomena, i.e. from an indirect observation. By level of integration, we consider whether the data relate to an individual or a group of individuals. Table 8.2 cross-references the nature and level of integration by following the data catalogue in Table 8.1.

The data used are mainly based on the direct observation of the phenomena in question. Working on mobility, it is logical that these data

26 In Table 8.1 in the appendices, the resolution of the municipality is taken as the average size of a municipality in the Gard, obtained by simply dividing the area of the Gard by the number of municipalities.

27 It would be important to also process the average frequency, e.g. weekly to annually, which is that of an individual level of organization adjusting daily activity and travel to changes in professional (school and work) or personal (leisure, holidays, and social networks) activities.

28 The spatial resolution of these data is ideally correlated: meter to hectometer for daily use and kilometer or common for life.

concern flows – physical flows such as rain, river flow, or road and human traffic such as daily or residential travel patterns. They concern topological information such as road cut-off points or the location of the building. We also consider that the data concerning regulatory tools are derived from the direct observation.

Indirect observations generally require a mathematical operation to transform ancillary observations into an observation of interest. Physical examples are radar measurement (an algorithm transforms a measurement of the electromagnetic reflectivity of raindrops into an intensity), road traffic reconstruction (a simulation transforms an origin/destination matrix of the population into a road trip), or move between municipalities (a simulation uses standard residential mobility profiles to reproduce movements between municipalities).

	Direct	Indirect
Individual	Individual travel profiles Mapping by topological survey (cadastral, road, hydrography, etc.) Traffic count at SIREDO stations Soil rainfall River flow at gauges Historical location of road cuts Feedback interview Characteristics of fatal accidents Building permit (application) Residential trajectories Tax profile of cadastral parcels	Image mapping Radar rainfall Occurrence of extreme events Reconstituted traffic at closure points Daily road travel
Global	Individual profiles concerning movement Communal population and its movements Regulatory tools (PLU, POS, etc.) Regulatory texts Typical residential migration profiles Number of inhabitants in flood-prone areas Number of activities and jobs in flood-prone areas	Submergence of the territory and roads Inter-municipal migration deducted from apparent balances and migration profiles

Table 8.2. *Key terms of the crossover between the direct or indirect nature and the level of individual or global integration of the data mentioned in Table 8.1 (the terms are taken from Table 5.1 in Appendix 1)*

By level of integration of observations, we distinguish data characterizing an individual from those characterizing a population, a group of individuals. Table 8.2 classifies a majority of data as individual in the sense that they are specific to a particular individual and, for example, their mobility – life trajectory or daily journey disrupted by a flood. Statistically characterized as standard travel profiles, these data become global in the sense that a profile is representative of a socio-demographic group. Following the same reasoning, the regulations that apply to a local or national population are also global.

For physical data, the distinction is more between being punctual and areal and may be open to discussion since the elementary unit is often defined arbitrarily. Rain, seen as a population of drops, is always approached in a global way because it is treated as a flux (intensity of the rain gauge) or a volume (radar reflectivity). On the other hand, if we consider that the sampling area of a rain gauge or radar measuring cell is very small compared to the surface of a watershed, we do have an individual point data. The elementary unit characterizing the river is the reach in which the "point" flow measurement is taken and in which the road cut is located. This reach is fed by a reach upstream and also by the slopes that build the river basin. On the other hand, classifying the flooding of land and roads as global in Table 8.2 is questionable, even though we have chosen this option. The rain-flow model used for the calculation of the submergence of the territory has a surface dimension and is therefore global, but the final data used remain point estimates of flows.

8.3.3. *Usefulness of reading by classes of data shared between being physical and human*

To sum up this somewhat academic exercise which sweeps away our problem with the help of large classes of data, leads to various points of clarification on the characteristic dimensions of the phenomena and their level of description.

Talking about data resolution leads to talking about the characteristic dimensions of the object or phenomenon being considered. Characterizing the dimensions of a phenomenon can be an objective in itself, especially when one wants to situate it in relation to others and understand their interactions. For continuous phenomena sampled quantitatively on a regular basis over time and/or space, the characteristic size is measurable by autocorrelation – a structural function that characterizes the average spatial–

temporal "pattern" of the phenomenon. This is the case for water flows, but also for road traffic, whose variability, without being easy to measure, is quite accessible. For these phenomena, the question arises of the correct adequacy between the sampling frequency and the characteristic dimension that leads to describing which scales of the phenomenon are resolved by the available data. For the other phenomena considered here, we do not know how to calculate a structure for different reasons. The sampling frequency imposed by the existing data can be too loose (census and transport surveys and cadastral plans) and leads to combining these data to perform "interpolation" (the use of standard profiles that contain the characteristic dimension). The phenomenon may not be easily quantifiable (individual trajectory and regulation). The characteristic time is then assessed in a more intuitive way, probably as the average time between changes in individual or collective behavior.

Talking about the nature and the level of data integration leads to questioning the level of description of phenomena and clarifying the mathematical transformations often implicitly used to develop the data. Accessing individual experience during a flood through the interview is clearly a direct observation – the witness is directly asked about his or her experience. If the physical information contained in the interview is used, the travel profile reconstructed from the story is a direct observation, if the good faith and the memory of the interviewee are accepted[29]. Once a level of interview coding has been used to segment an individual's behavior between states of information seeking, organization or protection against flooding, for example [CRE 09], the typical behavioral profiles thus identified are indirect observations.

8.4. Hierarchy of phenomena and mobilization of scales

The above-mentioned description exercise, combining phenomena and data, is a useful means of bringing together different scientific communities[30] and enabling them to share, intellectually if not in practice, a

29 Gaume *et al.* [GAU 04] show that the testimonies on the physical course of the floods are remarkably consistent and coherent with the measured quantities.

30 Even if a limit to this multidisciplinary working methodology can be set in the sense that it is directly inspired by the methodologies of the natural sciences and "imposed" as a framework for thinking in the human and social sciences. However, this limit is not specific to our working approach but is found in the majority of interdisciplinary projects. The complexity of environmental issues has boosted the momentum of interdisciplinarity,

very heterogeneous set of observation methods. It results in a description of the phenomena contributing to the exposure of populations to flash floods. We draw here some general considerations on the hierarchy of phenomena to which this exercise leads us and the scales or dimensions it has mobilized.

8.4.1. *A first level of decomposition in temporality and mobility*

The description proposed in Table 8.1 provides a first level of decomposition of the exposure process to flash floods using a dual dichotomy – that of the long term and the short term – that of water and individual mobility. This cross-over reveals a set of key phenomena that are distinctly physical or human, but also several phenomena that cross physical and human aspects[31] (see Table 8.3).

This first level of decomposition does not, therefore, deliver a simple juxtaposition of human and physical phenomena but also hybrids that are both human and physical, and as such call for data that are themselves mixed.

		Water mobility	Human mobility
Physical	Short term	Storm, flood, submersion	Road travel, decision
	Long term	Climate, extremes	Building deployment, residential migration, impact
Human	Short term	Alert, individual and collective protection	Time schedule
	Long term	Build dykes, dams	Life course, regulation

Table 8.3. *Key terms of temporality-mobility cross-overs when distinguishing between the physical and human dimensions. Note: terms in italics do not appear in Table 8.1 (in Appendix 1)*

reflecting a certain equivalence of knowledge, but the natural sciences still mainly lead the movement [ASP 12]. On the conditions for the development of interdisciplinary work, see Chapter 8 of this volume.

31 It is interesting to note, here, the possibility of mathematically representing the physical flows of a river and the human flows linked to road traffic by the same set of equations, because, in both cases, the velocity of the flow depends on its concentration [LIG 55a], [LIG 55b]. For this description, the problem of road flooding is critical because the speed–density relationship is the opposite for the river and traffic, positive for one, negative for the other – roads are congested where water speeds are highest.

This first level of deconstruction results in a collection of fairly heterogeneous phenomena. Some of them seem quite precisely circumscribed as physical phenomena but also as the daily program of individuals or the construction of buildings in relation to floodplains. Others, on the contrary, seem vast and need to be further broken down, such as the impact of remarkable events, urban planning, and risk regulations or the migration of individuals in their geographical, economic, and social contexts. Finally, others relating to human action on water mobility are rarely explored here (direct protective measures, such as the construction of dikes, for example). In other words, the effect of the flooding on the population with the causal sequence of phenomena ranging from storms, floods, and submergence of roads and buildings to the individual and institutional adaptation of the population in their daily and residential mobility is more detailed than the inverse effect of flood protection through warnings and containment or zoning[32].

8.4.2. *The convening of a broader set of dimensions*

While time and space, to which we return in the next paragraph, provide sufficient support for the physical description of the flood issue, it is in an expected way that the human description clearly requires the use of other dimensions.

Following Cash *et al.* [CAS 06], the administrative (from communal to national) and institutional (from rule to constitution) dimensions are useful to better describe the interplay of regulation and law hidden behind the "national regulation on urban planning and risk and its application at the local level" mentioned in Table 8.1. This process combines phenomena with

32 This bias should be addressed by studying the means of struggle in times of crises (e.g. inflatable dikes) and between crises (dikes and reservoirs). The policy of active protection against flash flooding is complex to implement (and undoubtedly very costly) in view of the dispersion of the risk along a very large river. Beyond the traditional individual protections blocking entrances, the ephemeral containment structures, present on the slow flooding market, are not to our knowledge used in the cases observed. A wide range of small-scale works are being carried out to control "ordinary" rainwater, particularly around roads and urban areas. As on a larger scale, this work creates the same perverse illusion of protection against "extraordinary" rains (effect observed, for example, during the accident at the Biot retirement home during a storm in September 2015). To our knowledge, historical examples of large-scale protection such as the Gard cliff dams built following the 1958 flood remain poorly studied in their positive or negative role in delaying the flooding of a few tributaries.

administrative and institutional dimensions such as the mobilization and the production of the norm by flooding: how at different administrative levels and through the interaction between these different levels, the phenomenon of flooding mobilizes the existing laws and regulations and promotes their evolution, i.e. acts on the institutional dimension?

Similarly, the managerial (tasks, project, and strategy) and network (individual, family, neighborhood, etc.) dimensions help to further analyze the "individual decision and adaptation to flooding" also mentioned in Table 8.1. This process is based on decision rules sought in interviews or expert reports. These rules use hierarchies at different managerial levels to manage the transition from routine to adaptive behavior: how will an individual strategy allow, as daily tasks are carried out, to switch from an ordinary activity project to a flood adaptation project? They also take into consideration the social constraints associated with embedding the individual in a complex social network: how are family and professional constraints integrated into the rules of adaptation?

It is important to note that the dimensions discussed here do not have a simple correlation with the spatial and temporal dimensions to which we return below. While the administrative and institutional levels have a fairly close link with the spatial dimensions but more precisely with the mass population dimensions [GRO 16], and the managerial and network dimensions have none. All these dimensions have no simple correlation with time: each level has several possible operating frequencies.

8.4.3. *Projection in time and space*

The temporal and spatial resolutions grouped in Table 8.1 locate the different dynamics of the phenomena considered in relation to each other, in terms of orders of magnitude[33]. The proposed projection in Figure 8.1 directly reflects all the information in Table 8.1.

This spatio-temporal representation makes it possible to identify the overlaps and specificities of the processes observed in the analysis. A first set of processes is characterized by a temporal variability from minute to day and a spatial variability of the precise address to the department. It concerns

33 We assimilate here the resolution and the characteristic dimension, which is a strong hypothesis if we refer to what is said in section 8.2.

meteorological processes of mesoscale convective systems, the submergence of the road network, or the territory as a whole that interact with daily activities and movements, requiring adaptation of these daily programs and decision-making both individually and collectively that must be taken quickly (on the order of the hour).

The second set of processes is observable at extended spatial and temporal resolutions: from a precise address at international level and from month to multi-year. The processes concerned relate to climate variability, on the one hand, and land use dynamics or territorial development choices, on the other hand, the interaction of which can lead to the implementation of legislative or regulatory measures, or even residential mobility choices in very specific cases (as shown in Chapter 3 of this volume).

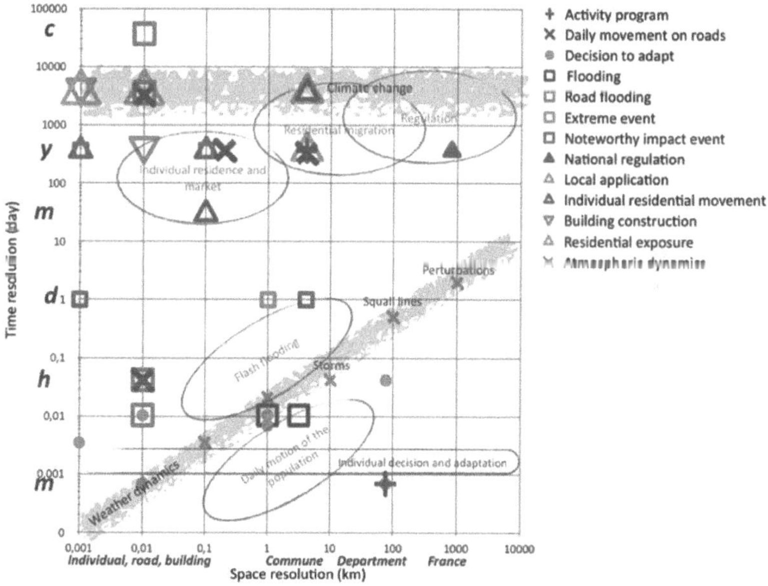

Figure 8.1. *Projection of the data described in Table 8.1 according to their resolution and representation of the estimated/assumed characteristic times. For a color version of this figure, see www.iste.co.uk/lutoff/hydrometeo2.zip*

8.5. Reflection on the method as a conclusion

This interdisciplinary observation of the confrontation between the dynamics of flash floods and social dynamics has thus enabled us to work on the elementary processes to be observed and on the units of measurement to be retained to enable an integrated analysis. It has led us to question what a phenomenon is and we have highlighted the difficulty of producing an observation that integrates the multiple scales and disciplines to be convened to define its goals.

To progress toward integration, we have tried to project our analysis into rich spaces such as those proposed by Cash *et al.* [CAS 06] or Grossetti [GRO 16]. However, while these references have provided food for thought, each of them imposes a drastic reduction in the number of dimensions observed, which complicates discussion between different disciplines, as the representatives of each discipline may be frustrated by the simplifications made in the analysis. However, the interdisciplinary work has also made us aware of the need to simplify the purpose of each discipline to allow integration with others.

Finally, the approach we followed in this observation and analysis can be summarized by the representation of Figure 8.2, inspired by the work of [ARE 02] on the modeling of individual activity programs. Depending on the characteristics of this model, which focuses on individuals, we propose here an approach to these processes considered at the individual level: what are the factors and parameters to be taken into account to understand the adaptation processes of an individual in the face of the potential or proven threat of flash flooding at different time scales? Nevertheless, we were able to test that this form of representation also makes it possible to define the processes at stake in terms of other social scales (as defined by [GRO 07]).

The center circle contains the parameters to be considered at the individual level according to different time steps considered in this adaptation process: real time of the event, medium term of the inter-crisis, and long term of the life courses. The bubbles that are distributed outside this central circle represent different categories of constraints to which the individual is subjected: climatic or meteorological (cyclical) constraints, environmental, socio-political constraints, and also the specific characteristics of the individual and their family environment or cognitive abilities. The interest of this representation is to summarize all the parameters and, therefore, the data to be considered in order to organize the socio-hydro-meteorological observation necessary to understand the processes involved at different time steps.

It should be noted, however, that the phenomena analyzed in this experiment are not exhaustive: for example, walking and outdoor activities are missing, which are not taken into account in the analysis we conducted. But, here, again, the interdisciplinary work carried out in the context of projects requires the selection of processes to be observed and simplifications. Nevertheless, we assume that the analytical approach proposed here must be able to accommodate other parameters and be used for application to other complex situations, as a result of the confrontation between physical and social dynamics.

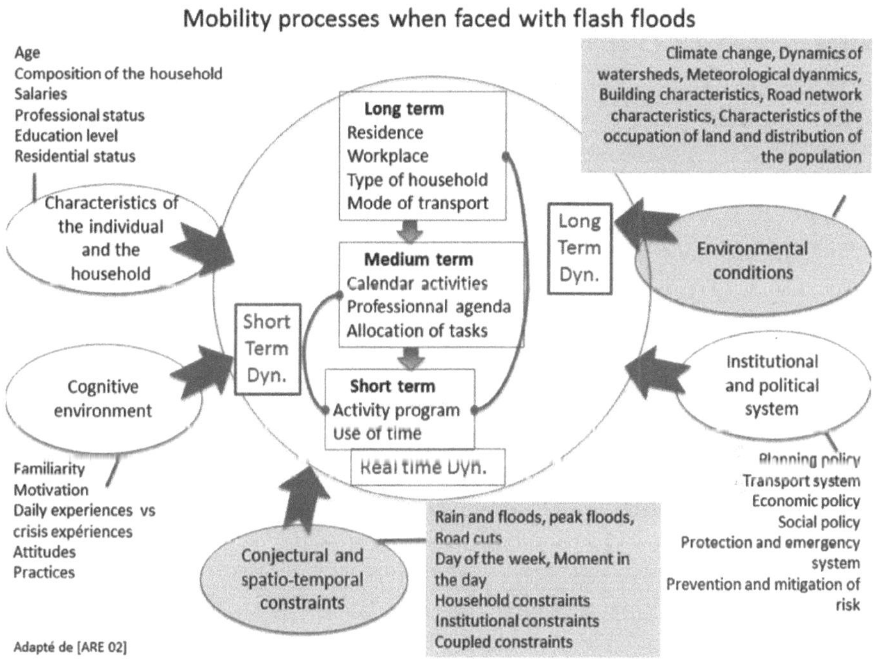

Figure 8.2. *Observation of social mobility in the face of flash flooding at different time scales. For a color version of this figure, see www.iste.co.uk/lutoff/hydrometeo2.zip*

The scientific approach aims to define the observations necessary for the analysis once the question you want to solve has been asked. However, the practice implemented in our interdisciplinary experiment shows the considerable help provided by the description of the data "classically" used (nature, resolution, and field) – because they are available – in the identification of phenomena of interest. We hope that this has shown that, on the basis of this initial heterogeneous data set and through an effort to categorize and interact, original, counter-intuitive ideas, results can emerge that can help to view the problem from a new perspective.

8.6. References

[ASP 12] ASPE C., JACQUÉ M., *Environnement et Société*, Quae/Fondation de la Maison des Sciences de l'Homme, Paris, 2012.

[ARE 02] ARENTZE T., TIMMERMANS H., "Modeling the formation of activity agendas using reactive agents", *Environment and Planning B: Planning and Design*, vol. 29, no. 5, pp. 719–728, 2002.

[BOC 13] BÖCKER L., DIJST M., PRILLWITZ J., "Impact of everyday weather on individual daily travel behaviours in perspective: a literature review", *Transport Reviews*, vol. 33, no. 1, pp. 71–91, 2013.

[BOR 08] BORGA M., GAUME E., CREUTIN J.D. *et al.*, "Surveying flash floods: gauging the ungauged extremes", *Hydrological Processes*, vol. 22, no. 18, 3883–3885, 2008.

[BRA 18] BRAUD I., VINCENDON B., ANQUETIN S. *et al.*, "The Challenges of Flash Flood Forecasting", *Mobility in the Face of Extreme Hydrometeorological Events 1*, ISTE Press Ltd, London and Elsevier Ltd, Oxford, 2018.

[CAS 06] CASH D., ADGER W., BERKES F. *et al.*, "Scale and cross-scale dynamics: governance and information in a multilevel world", *Ecology and Society*, vol. 11, no. 2, p. 8, 2006.

[CHA 74] CHAPIN F.S., *Human Activity Patterns in the City: Things People do in Time and in Space*, vol. 13, Wiley-Interscience, New York, 1974.

[COO 13] COOLS M., CREEMERS L., "The dual role of weather forecasts on changes in activity-travel behavior", *Journal of Transport Geography*, vol. 28, pp. 167–175, 2013.

[CRE 09] CREUTIN J.D., BORGA M., LUTOFF C. *et al.*, "Catchment dynamics and social response during flash floods: the potential of radar rainfall monitoring for warning procedures", *Meteorological Applications*, vol. 16, pp. 115–125, 2009.

[CRO 16] CROMMELINCK S., BENNETT R., GERKE M. *et al.*, "Review of automatic feature extraction from high-resolution optical sensor data for UAV-based cadastral mapping", *Remote Sensing*, vol. 8, p. 689, 2016.

[FOU 12] FOUILLÉ L., BROC J.S., BOURGES B. *et al.*, "La place des modèles de trafic dans les récentes modélisations des impacts environnementaux des transports. Importance de l'explicitation des méthodes et hypothèses", *Recherche Transports Sécurité*, vol. 28, nos 3–4, pp. 190–200, 2012.

[GAU 04] GAUME E., LIVET M., DESBORDES M. *et al.*, "Hydrological analysis of the river Aude, France, flash flood on 12 and 13 November 1999", *Journal of Hydrology*, vol. 286, no. 1, pp. 135–154, 2004.

[GRO 16] GROSSETTI M, "L'imprévisibilité dans le monde social", in LEVY S. (ed.), *Complexité et désordre. Éléments de réflexion*, Presses Universitaires de Grenoble, Grenoble, pp. 97–112, 2016.

[HAG 70] HAGERSTRAAND T., "What about people in regional science?", *Regional Science*, vol. 24, no. 1, pp. 7–24, 1970.

[KAU 98] KAUFMANN J., STEUDLER D., *Cadastre 2014: A Vision for a Future Cadastral System*, Publication of Fédération Internationale des Géomètres (FIG), Commission 7-Cadastre and Land Management, Brighton, England, 1998.

[KIL 07] KILPELÄINEN M., SUMMALA H., "Effects of weather and weather forecasts on driver behaviour", *Transportation Research Part F: Traffic Psychology and Behaviour*, vol. 10, no. 4, pp. 288–299, 2007.

[LIG 55a] LIGHTHILL M., WHITHAM G., "On kinematic waves. I. Flood movement in long rivers", *Proceedings of the Royal Society of London. Series A. Mathematical and Physical Sciences*, vol. 229, p. 281, 1955.

[LIG 55b] LIGHTHILL M., WHITHAM G., "On kinematic waves. II. A theory of traffic flow on long crowded roads", *Proceedings of the Royal Society of London. Series A. Mathematical and Physical Sciences*, vol. 229, p. 317, 1955.

[NAU 12] NAULIN J.-P., Modélisation hydrologique distribuée pour la prévision des coupures de routes par inondation: application au département du Gard, PhD thesis, Ecole centrale de Nantes, Nantes, 2012.

[SHA 16] SHABOU M.S., Extrêmes hydrométéorologiques et exposition sur les routes. Contribution à MobRISK: modèle de simulation de l'exposition des mobilités quotidiennes aux crues rapides, Earth Sciences PhD thesis, Université Grenoble Alpes, Grenoble, 2016.

[VER 07] VERSINI P.A., Surveillance du risque hydrologique diffus le long des itinéraires routiers: Eléments pour la construction d'un outil d'annonce de coupure, PhD thesis, Ecole des Ponts ParisTech, Paris, 2007.

Conclusion

This second volume concludes the overview we offer with this book on interdisciplinary research work undertaken over the past 15 years to improve understanding of the interactions between the dynamics of flash flood phenomena and the dynamics of social mobility.

In the first volume [LUT 18], we hypothesized the heuristic value of the notion of scale to address the complexity of flash floods, phenomena that involve multiple spatial and temporal scales. To understand the interactions between these extreme events and social dynamics requires working both at very narrow scales (short time of the crisis, on limited surfaces) and at very broad scales (long time of the evolution of territories, of the evolution of social norms, of climate dynamics). A more precise interest in the social processes involved in these events still requires us to cross-reference the views of different social scales, from the individual actor (from behavior during the crisis to residential choices that both influence exposure) to the State (or even international organizations) and human collectives of varying geometry. This first work thus made it possible to identify the scales relevant for the analysis of interactions between flash floods and social mobility, but also showed the need to create new reference scales to address certain points identified as major ones. This concerns, in particular, the construction of a scale of event impacts that would make it possible to go beyond the single flow forecast, by integrating the possible impacts on societies. Some avenues have been opened up – in particular, the construction of a severity scale that still requires specific research developments.

Conclusion written by Séverine DURAND and Céline LUTOFF.

Reflection on the notion of scale has made it possible to identify another avenue of research: that of the levels of adaptation or resistance that can be observed according to the contexts and the levels of action and decision considered. We thus concluded in the first volume on the need to progress on the analysis of these postures and attitudes and, in particular, to question the paces and cycles of adaptation.

It is to this task that we have set out in this second volume by questioning how extreme events can stimulate profound changes in social dynamics and pace adaptation in a specific way at each scale (temporal, spatial, social, and institutional).

The challenge of this second volume was, thus, to understand what mobilizes the social response, on the one hand, and what drives its pace on the other hand. One of the central questions was to identify the elements that enable the transition from one reference model to another. How many events are required and on what time scale to implement risk management measures at the local level or more broadly? Another central question concerns, for example, the scales of individual decisions: what are the relevant environmental indicators that make individuals switch from their daily activities to behaviors involving anticipating the event and protection?

Before addressing these questions, it seemed important to us to frame the work presented in the interdisciplinary approach that characterizes them. "The interdisciplinary journey" proposed in the first chapter thus aims to draw lessons from our experience and to propose to the reader to enter into this book by reflecting on what this approach allows and implies. This is an important subject in the context of significant development of interdisciplinary work, the most appropriate to grasp the complexity of contemporary environmental issues. Concerning the adventure that brought us together during the work presented in this book, Chapter 1 shows that this interdisciplinary experience has made it possible to progress from a thematic point of view on the understanding of the socio-hydrometeorological processes at play during flash flood events. More broadly, this has made it possible to develop methodological tools that facilitate an interdisciplinary approach to environmental issues. These are tools that are now being mobilized in the context of other issues such as climate change and its impacts on socio-ecosystems, also involving other disciplines[1].

Once the framework on our interdisciplinary approach was established, we approached the question of adaptation rates throughout the seven chapters which followed, differentiating between the specific tempos of the crisis (short-term analysis) and those of the inter-crisis and prevention (long-term analysis).

Chapters 2 and 3 were devoted to long-term analysis. Chapter 2 dealt with the issue of urbanization rates in flood-prone areas. Applied to a case study at a departmental scale (Gard), the analysis shows that the occurrence of extreme hydrometeorological events produces short-term breaks in the rates of urbanization observed over the long term. While a general decline in construction in flood-prone areas can be observed between 1950 and 2012 throughout the Gard, significant differences are visible, depending on the sector and the level of land pressure. At the municipal level, for example, significant declines are observed mainly in 2–3 years following the phenomenon. This change is also extremely variable depending on the size of the municipalities. This study thus seems to indicate a significant variation in land-use planning rates in relation to extreme hydrometeorological events.

Chapter 3 agrees in the same way by focusing the analysis on residential choices in high-risk areas. Indeed, the latter shows in support of a qualitative study among more than 50 inhabitants and real estate professionals that the risk of flooding does not seem to significantly influence the choice to settle or remain on a floodplain. The analysis shows that the conceptualization of residential choices through the search for objective attributes is not very relevant and cannot express all the preferences that nourish them. The authors thus conclude on the interest of studying situations "in the process of being done" by focusing attention on actors' practices in situations, taking seriously the attachments that link them to their living environments and the environmental entities that constitute them, to better understand social complexities.

Chapters 4 and 5 focus on the adaptation rates at the time of the event by studying the specific exposure of motorists in times of crisis.

1 In particular, ecologists within the framework of the *Trajectories* project at the Université Grenoble Alpes, see Chapter 1 of this Volume.

Chapter 4 provides a model of this exposure tested with the data of a real case. This work has shown that the exposure index varies significantly according to the status and professional activity of individuals, their age and gender, with middle-aged men being the most exposed. Nevertheless, the model can be refined to improve the consideration of the modeled processes and to optimize the calculation time, making it more operational. These synthetic elements thus encourage further work on this modeling of human exposure to flash floods.

Chapter 5 provides some answers to this question in order to improve the analysis of daily mobility, by using an analysis of individuals' activity programs to refine the understanding of daily social rhythms and the place occupied by travel within them.

This work thus reveals the potential of these tools, which model the adaptation rates at the time of the event for crisis management and emergency action planning, and also for road management – in particular, to locate in space and time the most dangerous road/river intersections due to the number of individuals likely to cross them at the time of submergence.

The last three chapters of the book dealt more specifically with methodological aspects of the analyses proposed in this book on adaptation paces.

Chapter 6 explored the benefits of using data from social networks as a tool for observing social response patterns in the occurrence of an event. The observed case is based on the data from the Twitter platform, allowing access to geolocated tweets. The authors concluded on the need to qualify the ability of tweets to serve as a material for the construction of objective knowledge by showing the limits of their analysis. For time being, geolocated tweets – although not without interest on certain points – do not make it possible to reconstruct, in an exhaustive way, the complexity of a hydro-meteorological crisis.

To better understand the social paces of adaptation of social mobilities – both throughout the crisis and over the long-term in terms of prevention – Chapter 7 reports on a *physico-social* study that proposes a centered (on the peak flood) and reduced (by the response time of watersheds over the short term and by the recurrence of floods over the long term) time scale that makes it possible to jointly analyze physical and social cursors to better understand the dynamics at work during the crisis and between crises. From the point of view of the social sciences, this standardization of observations by physics allows a shift in reference factors that renews the analysis of social responses to flooding (usually based on social categorizations). From a climate science perspective, the power of weather forecasting to normalize social observations is questioned when the current scale at which this forecast is provided (departmental) is not sufficient to understand the actions undertaken more locally by individuals. The authors invite others to continue the work using this socio-hydro-meteorological approach to study, for example, how flood markers could be mobilized over the time of the inter-crisis to study the pace of events at different spatial scales and to understand how these events could encourage (or not) preventive responses through the different interlocking scales (locally, regionally, nationally).

Finally, Chapter 8 provided an overview of all the data that had to be collected and integrated collectively and in an interdisciplinary approach to address our question of the interaction between the dynamics of flash floods and social dynamics. It shows, in particular, the considerable help provided by the description of the data "classically" used because they are available (according to their nature, resolution, field) in the identification of phenomena of interest. It shows that on the basis of an *a priori* heterogeneous dataset and through an effort to categorize and interact, original, counter-intuitive ideas and results can emerge that can help to renew perspectives on the objects studied.

Whether through the prism of scales (Volume 1) or paces (Volume 2), all the works presented in this book will, we hope, have contributed to advancing knowledge of extreme socio-hydro-meteorological events. Some of the research avenues they have opened up and some of the work and developments they have called for are already being explored. In the continuity of the work presented, the ambition remains intact: that of an interdisciplinary adventure to improve societies' ability to anticipate and adapt to climatic extremes forecasted to multiply in the context of climate change that we are already experiencing.

References

[LUT 18] LUTOFF C., DURAND S., *Mobility in the Face of Extreme Hydrometeorological Events 1: Defining the Relevant Scales of Analysis*, ISTE, London, 2018.

Appendix 1: Tables for Chapter 5

	Operations used	
Distance metrics	**Substitution**	**Insertion and deletion**
Hamming	Yes (cost = 1)	No
Levenshtein I (Optimal Matching)	Yes (cost = 1)	Yes (cost = 1)
Levenshtein II	No	Yes (cost = 1)

Table 5.1. *Different methods for computing distances between sequences. The cost parameters refer to the default weight associated with each operation*

Variables	**Categories**	**Percent (%)**
Gender	Male	44.3
	Female	55.7
Age	<18 years old	7.55
	18 – 29 years old	13.43
	30–45 years old	25.35
	46–60 years old	24.99
	>60 years old	28.68
Education level	No diploma	35.18
	School–College	19.89
	Bachelor	35.18
	University	25.35

Variables	Categories	Percent (%)
Profession	Farmers	1.09
	Artisans – Trades people	3.26
	Managerial staff	8.76
	Laborer	13.47
	Intermediate professions (professors)	15.45
	Employee	10.12
	Retired	29.61
	Unemployed	12.04
	Non-administrated	6.21
Professional status	Working	46.69
	Student	5.33
	Unemployed	4.83
	Retired	28.47
	Househusband/wife	5.21
	Other situation	2.92
	Non-administrated	6.54
Size of household	1 person	31.32
	2 persons	33.89
	>2 persons	34.8
Household composition	Single	31.32
	One parent family	8.07
	Couple with/without children	57.41
	Other composition	3.20
Number of cars by household	No car	15.60
	1 car	48.43
	>1 car	35.96

Table 5.2. *Socio-demographic description of the study area population*

Variables	Categories	Average trip number by category for 1 weekday	Average trip number by category for 1 weekend
Gender	Male	3.61	**3.23**
	Female	**3.76**	3.17
Age	<18 years old	3.40	2.98
	18–29 years old	3.83	3.28
	30–45 years old	**4.35**	**3.49**
	46–60 years old	3.8	3.30
	>60 years old	3.18	2.89
Education level	No diploma	3.39	2.98
	School – College	3.3	2.98
	Bachelor	3.86	3.27
	University	**4.04**	**3.49**
Professional status	Working	4.04	**3.46**
	Student	3.44	3.21
	Unemployed	**4.08**	3.09
	Retired	3.25	2.91
	Househusband/wife	3.91	3.04
	Other situation	3.35	2.99
Size of household	1 person	3.37	3.04
	2 persons	3.46	3.12
	>2 persons	**3.79**	**3.37**
Household composition	Single	3.37	3.04
	One parent family	**4.08**	3.30
	Couple with/without children	3.79	**3.25**
	Other composition	3.6	3.08
Number of cars by household	No car	3.9	2.84
	1 car	3.07	3.16
	>1 car	**3.66**	**3.32**
Total population		3.69	3.19

Table 5.3. *Variation of average trip number by socio-demographic category in weekday and weekend (the numbers in bold refer to the category with the highest average number of trips)*

	Types of activities			
	Travel	**Home**	**Mandatory**	**Secondary**
Travel	0	1	1	1
Home	1	0	1	1
Mandatory	1	1	0	1
Secondary	1	1	1	0

(a)

	Types of activities			
	Travel	**Home**	**Mandatory**	**Secondary**
Travel	0	3	3	3
Home	3	0	2	2
Mandatory	3	2	1 (0 when it is the same activity)	2
Secondary	3	2	2	1 (0 when it is the same activity)

(b)

Table 5.4. *Two parameterizations of cost setting of Optimal Matching metric. Activities are classified into four principal categories: travel, in-home activities, out-of-home mandatory activities (work and school), and out-of-home secondary activities (shopping, administrative, medical, accompanying and leisure activities). For the first cost setting (Table 5.4.a), an equivalent substitution cost (c = 1) is implemented. In the second cost setting (Table 5.4.b), substitutions costs are set with respect to the activity category. We accentuated the difference between travel activities and the other categories (c = 3), in-home activities and out-of-home activities (c = 2) and between mandatory and secondary activities (c = 2)*

Type of day	**Variables**	***F***	R^2	***p*-value**
Average week day	Age	16,596.38	**0.102**	0.001*
	Gender	1,882.29	0.003	0.019*
	Education level	6,198.73	0.030	0.001*
	Professional status	17,347.59	0.151	0.001*
	Profession	10,890.80	**0.129**	0.001*
	Household size	4,551.43	0.015	0.001*

Type of day	Variables	*F*	R^2	*p*-value
	Number of children	5,305.90	0.017	0.001*
	Household type	1,283.36	0.006	0.048
	Number of cars	4,354.07	0.014	0.001*
Global		8.47	0.237	0.001*
Average weekend day	Age	3,059.15	**0.023**	0.001*
	Gender	1,394.32	0.002	0.086
	Education level	1,260.22	0.007	0.027*
	Professional status	1,929.23	**0.022**	0.001*
	Profession	1,797.96	**0.027**	0.001*
	Household size	1,484.38	0.005	0.016*
	Number of children	1,192.81	0.004	0.089
	Household type	1,146.78	0.006	0.104
	Number of cars	1,512.89	0.005	0.015*
Global		1.86	0.073	0.001*

Table 5.5. *Results of the discrepancy analysis of activities sequences for an average weekday schedule and an average weekend schedule with respect to the default cost setting (Table 5.4.a). The statistics F and R^2 are computed as shown in equations [5.3] and [5.4]. The p-value represents the statistical significance of the effect of each variable. It is generally admitted that 5,000 permutations should be used to assess a significance threshold of 1% and 1,000 for a threshold of 5% [MAN 06]*

Type of day	Variables	*F*	R^2	*p*-value
Average week day	Age	20,484	**0.123**	0.001*
	Gender	1,946	0.003	0.016*
	Education level	7,243	0.035	0.001*
	Professional status	20,039	**0.170**	0.001*
	Profession	12,142	**0.142**	0.001*

Type of day	Variables	*F*	R^2	*p*-value
	Household size	4,551	0.015	0.001*
	Number of children	6,728	0.022	0.001*
	Household type	1,279	0.006	0.064
	Number of cars	4,404	0.014	0.001*
Global		9.98	0.268	0.001*
Average weekend day	Age	3,977	**0.030**	0.001*
	Gender	1,509	0.002	0.048*
	Education level	1,421	0.008	0.031*
	Professional status	2,252	**0.026**	0.001*
	Profession	2,041	**0.031**	0.001*
	Household size	1,825	0.007	0.006*
	Number of children	1,353	0.005	0.060
	Household type	1,248	0.007	0.088
	Number of cars	1,664	0.006	0.013*
Global		2.26	0.088	0.001*

Table 5.6. *Results of the discrepancy analysis of activities sequences for an average weekday schedule and an average weekend schedule with respect to the customized cost setting (Table 5.4.b). The statistics F and R^2 are computed as shown in equations [5.3] and [5.4]. The p-value represents the statistical significance of the effect of each variable. It is generally admitted that 5,000 permutations should be used to assess a significance threshold of 1% and 1,000 for a threshold of 5% [MAN 06]*

Appendix 2: Table 1 in Chapter 8

Phenomenon	Nature of the data	Data source	Resolution		
			Time (day)	Space (km)	Time
Daily travel program for the entire population	Typical travel profiles (length, duration, mode, time) by socio-demographic category	http://www.statistiques.developpement-durable.gouv.fr/accueil.html then ENTD 2008	6.94E-04	76,505	Min
	Combination of "type" profiles according to the socio-demographic composition of the population	https://www.insee.fr/fr/information/2409289 on 2013 Census	3.65E+02	4,072	Year
Daily road travel of the entire population	Origin-destination matrices for professionals and schools by the municipality	https://www.insee.fr/fr/statistiques/2022115 on MobPro and MobSco data	3.65E+02	4,072	Year
	Cadastral information used to locate tax revenues within municipalities	https://www.insee.fr/fr/information/1303416 on Localized tax revenues	3.65E+02	0.200	Year

Phenomenon	Nature of the data	Data source	Resolution		
			Time (day)	Space (km)	Time
	Geometry and technical characteristics of the road network	BD CARTO® http://professionnels.ign.fr/	3.65E+03	0.010	Decade
	Road traffic characteristics (flow, rates, speeds) in a set of counting stations	SIREDO system http://trafic-routier.data.cerema.fr/siredo-systeme-national-de-recueil-de-donnees-de-a37.html for the Gard department	4.17E-02	0.010	Hour
Submergence of the territory by surface runoff owing to thunderstorms	Rain intensity	Météo France and OHMCV http://www.ohmcv.fr	1.04E-02	1,000	Quarter-hour
	River flows measured at gauging stations and estimated by rain-flow modeling at all points in the hydrographic network	SEVNoL® BD and OHMCV CVN model http://www.ohmcv.fr	1.04E-02	3,162	Quarter-hour

Phenomenon	Nature of the data	Data source	Resolution		
			Time (day)	Space (km)	Time
Submergence of roads during stormy events	Historical location of road network cut-off points due to flooding and their sensitivity	BD PITCH® and BD road cuts of the Gard County Council	4.17E-02	0.010	Hour
	Geometry of the hydrographic network	http://www.statistiques.developpement-durable.gouv.fr/accueil.html then ENTD 2008	3.65E+04	0.010	Century
	Geometry and technical characteristics of the road network	BD CARTO® http://professionnels.ign.fr/	3.65E+03	0.010	Decade
	Probabilities of flooding at cut-off points by hydrometeorological event	IFSTTAR analysis of PITCH data and IGE rain-flow simulations CVN model	1.04E-02	0.010	Quarter-hour

Phenomenon	Nature of the data	Data source	Resolution		
			Time (day)	Space (km)	Time
Decision and individual adaptation to the flood	Feedback interviews: corpus of texts	Feedback interviews – OHMCV-HymEx	3.47E-03	0.001	5 mins
	Precipitation intensities	Weather France	1.04E-02	1,000	Quarter hour
	Intensities related to time, season and rain	OHMCV Re-analysis http://www.ohmcv.fr and day length model	6.94E-03	1,000	10 mins
	Vigilance level (amber, red)	Weather France	4.17E-02	76,505	Hour
	Typical travel profiles (length, duration, mode, time) by socio-demographic category	http://www.statistiques.developpement-durable.gouv.fr/accueil.html then ENTD 2008	6.94E-04	76,505	Min
	Probabilities of flooding at cut-off points by hydrometeorological event	IFSTTAR analysis of PITCH data and IGE rain-flow simulations CVN model	1.04E-02	0.010	Quarter-hour

Phenomenon	Nature of the data	Data source	Resolution		
			Time (day)	Space (km)	Time
	Characteristics of fatal accidents: location, time, circumstances, victim characteristics	OHMCV http://www.ohmcv.fr	6.94E-04	0.010	Min
Occurrence of extreme events	Precipitation intensities adjusted to statistical laws	Météo France and OHMCV http://www.ohmcv.fr	1.00E+00	1,000	Day
Impacts of significant events	Data catalogs of remarkable episodes		1.00E+00	4,072	Day
	Local regulatory tools for political, legal and administrative land use management: PLU and PLUI	http://www.cohesion-territoires.gouv.fr/plan-local-d-urbanisme-intercommunal-plui-et-plan-local-d-urbanisme-plu	3.65E+02	0.100	Year
	Local regulatory tools for risk management: PPRIs and similar	http://www.gard.gouv.fr/Politiques-publiques/Securite-et-protection-de-la-population/Risques/Gestion-du-risque-inondation/Plasns-de-Prevention-des-Risques-Inondation-PPRI	3.65E+02	0.001	Year

Phenomenon	Nature of the data	Data source	Resolution		
			Time (day)	Space (km)	Time
	Location and characteristics of building permits issued		1.00E+00	0.001	Day
National urban planning and risk regulations and their application at the local level	Texts of implementing laws and decrees and environmental code	https://www.legifrance.gouv.fr/	3.65E+02	802,372	Year
	Local regulatory tools for risk management: PPR, article R111-3	https://www.collectivites-locales.gouv.fr/lenvironnement	3.65E+03	0.010	Decade
	Local regulatory tools for urban planning: PLU, POS, SCOT	https://www.collectivites-locales.gouv.fr/documents-durbanisme-et-regles-generales-durbanisme	3.65E+03	0.010	Decade
	Building permits		3.65E+03	0.001	Decade

Phenomenon	Nature of the data	Data source	Resolution		
			Time (day)	Space (km)	Time
Residential migration of individuals in their geographical, economic and social context	Individual narratives of residential trajectories	Interview field survey	3.00E+01	0.100	Month
	Expert discourses on individual trajectories and motivations	Interviews with real estate agents, insurers, lawyers, urban planners	3.65E+02	0.100	Year
	Typical individual migration profiles linking the situation of municipalities, distance, and socio-demographic categories	Quetelet Network www.reseau-quetelet.cnrs.fr and General Population Census INSEE https://www.insee.fr	3.65E+03	4,072	Decade
		BD TOPO and ALTI® http://professionnels.ign.fr/	3.65E+02	0.001	Year
Building works in relation to floodable areas	Location of buildings constructed outside and within flood risk areas based on a comparison of successive maps of the building	BD TOPO® http://professionnels.ign.fr/ and PPRI from DDTM Gard	3.65E+03	0.001	Decade

Phenomenon	Nature of the data	Data source	Resolution		
			Time (day)	Space (km)	Time
	Declared tax profile of cadastral parcels (type of buildings, owner and property, date of building permit, address)	DGFIP's MAJIC C land file https://www.collectivites-locales.gouv.fr/fourniture-fichiers-informatises-fonciers and PPRI, PLU, etc.	3.65E+02	0.010	Year
Migration and departmental residential exposure	Number of inhabitants and built areas outside and in flood risk areas according to land declarations	MAJIC land file processed by http://www.noe.gard.fr	3.65E+02	4,072	Year
	Number of activities and jobs in floodplain areas outside Nîmes	MAJIC land file processed by http://www.noe.gard.fr	3.65E+02	4,072	Year
	Inter-municipal migration derived from apparent migration balances and standard individual migration profiles	General Population Census INSEE https://www.insee.fr	3.65E+03	4,072	Decade

Table 8.1. *Observed phenomena, nature of data, sources and resolution (spatial and temporal)*

List of Authors

Isabelle ANDRÉ-POYAUD
UMR PACTE
Université Grenoble Alpes
France

Sandrine ANQUETIN
IGE
Université Grenoble Alpes
France

Brice BOUDEVILLAIN
IGE
Université Grenoble Alpes
France

Joël BOULIER
Géographie-cités
Panthéon Sorbonne Université
Paris
France

Isabelle BRAUD
HHLY – Irstea
Lyon Villeurbanne Centre
France

Camille CAVALIÈRE
INP
Université Grenoble Alpes
France

Sonia CHARDONNEL
UMR PACTE
Université Grenoble Alpes
France

Charles-Antoine CHOQUET
UMR PACTE
Université Grenoble Alpes
France

Jean-Dominique CREUTIN
IGE
Université Grenoble Alpes
France

Paule-Annick DAVOINE
UMR PACTE
Université Grenoble Alpes
France

Samuel DEBIONNE
IGE
Université Grenoble Alpes
France

Edwige DUBOS-PAILLARD
Géographie-cités
Panthéon Sorbonne Université
Paris
France

Véronique DUCROCQ
CNRM France
Toulouse
France

Séverine DURAND
UMR PACTE
Université Grenoble Alpes
France

Sylvie DUVILLARD
UMR PACTE
Université Grenoble Alpes
France

Céline LUTOFF
UMR PACTE
Université Grenoble Alpes
France

Antonin PAVARD
Géographie-cités
Panthéon Sorbonne Université
Paris
France

Isabelle RUIN
IGE
Université Grenoble Alpes
France

Valérie SANSEVERINO-GODFRIN
Mines ParisTech
France

Saif SHABOU
UMR PACTE
Université Grenoble Alpes
France

Béatrice VINCENDON
CNRM France
Toulouse
France

Index

P, R, S

T, U, V

W, Z